Selected Titles in This Series

D1786601

736 **Wojciech Chachólski and Jérôme Scherer,** Homotopy theory of diagrams, 2002
735 **Martina Brück, Xi Du, Joonsang Park, and Chuu-Lian Terng,** The submanifold geometries associated to Grassmannian systems, 2002
734 **Michel Van den Bergh,** Blowing up of non-commutative smooth surfaces, 2001
733 **Milé Krajčevski,** Tilings of the plane, hyperbolic groups and small cancellation conditions, 2001
732 **Jan O. Kleppe, Juan C. Migliore, Rosa Miró-Roig, Uwe Nagel, and Chris Peterson,** Gorenstein liaison, complete intersection liaison invariants and unobstructedness, 2001
731 **Jesús Bastero, Mario Milman, and Francisco J. Ruiz,** On the connection between weighted norm inequalities, commutators and real interpolation, 2001
730 **Suhyoung Choi,** The decomposition and classification of radiant affine 3-manifolds, 2001
729 **Michael Grosser, Eva Farkas, Michael Kunzinger, and Roland Steinbauer,** On the foundations of nonlinear generalized functions I and II, 2001
728 **Laura Smithies,** Equivariant analytic localization of group representations, 2001
727 **Anthony D. Blaom,** A geometric setting for Hamiltonian perturbation theory, 2001
726 **Victor L. Shapiro,** Singular quasilinearity and higher eigenvalues, 2001
725 **Jean-Pierre Rosay and Edgar Lee Stout,** Strong boundary values, analytic functionals, and nonlinear Paley-Wiener theory, 2001
724 **Lisa Carbone,** Non-uniform lattices on uniform trees, 2001
723 **Deborah M. King and John B. Strantzen,** Maximum entropy of cycles of even period, 2001
722 **Hernán Cendra, Jerrold E. Marsden, and Tudor S. Ratiu,** Lagrangian reduction by stages, 2001
721 **Ingrid C. Bauer,** Surfaces with $K^2 = 7$ and $p_g = 4$, 2001
720 **Palle E. T. Jorgensen,** Ruelle operators: Functions which are harmonic with respect to a transfer operator, 2001
719 **Steve Hofmann and John L. Lewis,** The Dirichlet problem for parabolic operators with singular drift terms, 2001
718 **Bernhard Lani-Wayda,** Wandering solutions of delay equations with sine-like feedback, 2001
717 **Ron Brown,** Frobenius groups and classical maximal orders, 2001
716 **John H. Palmieri,** Stable homotopy over the Steenrod algebra, 2001
715 **W. N. Everitt and L. Markus,** Multi-interval linear ordinary boundary value problems and complex symplectic algebra, 2001
714 **Earl Berkson, Jean Bourgain, and Aleksander Pełczynski,** Canonical Sobolev projections of weak type $(1,1)$, 2001
713 **Dorina Mitrea, Marius Mitrea, and Michael Taylor,** Layer potentials, the Hodge Laplacian, and global boundary problems in nonsmooth Riemannian manifolds, 2001
712 **Raúl E. Curto and Woo Young Lee,** Joint hyponormality of Toeplitz pairs, 2001
711 **V. G. Kac, C. Martinez, and E. Zelmanov,** Graded simple Jordan superalgebras of growth one, 2001
710 **Brian Marcus and Selim Tuncel,** Resolving Markov chains onto Bernoulli shifts via positive polynomials, 2001
709 **B. V. Rajarama Bhat,** Cocylces of CCR flows, 2001
708 **William M. Kantor and Ákos Seress,** Black box classical groups, 2001
707 **Henning Krause,** The spectrum of a module category, 2001
706 **Jonathan Brundan, Richard Dipper, and Alexander Kleshchev,** Quantum Linear groups and representations of $GL_n(\mathbb{F}_q)$, 2001

(Continued in the back of this publication)

Homotopy Theory of Diagrams

MEMOIRS
of the
American Mathematical Society

Number 736

Homotopy Theory of Diagrams

Wojciech Chachólski
Jérôme Scherer

January 2002 • Volume 155 • Number 736 (second of 5 numbers) • ISSN 0065-9266

American Mathematical Society
Providence, Rhode Island

2000 *Mathematics Subject Classification*.
Primary 55U35, 18G55; Secondary 18G10, 18F05, 55U30, 55P65.

Library of Congress Cataloging-in-Publication Data

Chachólski, Wojciech, 1968–
Homotopy theory of diagrams / Wojciech Chachólski, Jérôme Scherer.
p. cm. — (Memoirs of the American Mathematical Society, ISSN 0065-9266 ; no. 736)
Includes bibliographical references and index.
ISBN 0-8218-2759-6 (alk. paper)
1. Homotopy theory. 2. Categories (Mathematics) I. Scherer, Jérôme, 1969– II. Series.

QA3 A57 no. 736
[QA612.7]
510 s—dc21
[514′.24] 2001045783

Memoirs of the American Mathematical Society

This journal is devoted entirely to research in pure and applied mathematics.

Subscription information. The 2002 subscription begins with volume 155 and consists of six mailings, each containing one or more numbers. Subscription prices for 2002 are $524 list, $419 institutional member. A late charge of 10% of the subscription price will be imposed on orders received from nonmembers after January 1 of the subscription year. Subscribers outside the United States and India must pay a postage surcharge of $31; subscribers in India must pay a postage surcharge of $43. Expedited delivery to destinations in North America $35; elsewhere $130. Each number may be ordered separately; *please specify number* when ordering an individual number. For prices and titles of recently released numbers, see the New Publications sections of the *Notices of the American Mathematical Society.*

Back number information. For back issues see the *AMS Catalog of Publications.*

Subscriptions and orders should be addressed to the American Mathematical Society, P. O. Box 845904, Boston, MA 02284-5904. *All orders must be accompanied by payment.* Other correspondence should be addressed to Box 6248, Providence, RI 02940-6248.

Memoirs of the American Mathematical Society is published bimonthly (each volume consisting usually of more than one number) by the American Mathematical Society at 201 Charles Street, Providence, RI 02904-2294. Periodicals postage paid at Providence, RI. Postmaster: Send address changes to Memoirs, American Mathematical Society, P. O. Box 6248, Providence, RI 02940-6248.

This publication is indexed in *Science Citation Index®*, *SciSearch®*, *Research Alert®*, *CompuMath Citation Index®*, *Current Contents®/Physical, Chemical & Earth Sciences.*
Printed in the United States of America.

♾ The paper used in this book is acid-free and falls within the guidelines established to ensure permanence and durability.
Visit the AMS home page at URL: `http://www.ams.org/`

10 9 8 7 6 5 4 3 2 1 07 06 05 04 03 02

Contents

Abstract

In this paper we develop homotopy theoretical methods for studying diagrams. In particular we explain how to construct homotopy colimits and limits in an arbitrary model category. The key concept we introduce is that of a model approximation. A model approximation of a category $\mathcal{C}$ with a given class of weak equivalences is a model category $\mathcal{M}$ together with a pair of adjoint functors $\mathcal{M} \rightleftarrows \mathcal{C}$ which satisfy certain properties. Our key result says that if $\mathcal{C}$ admits a model approximation then so does the functor category $Fun(I, \mathcal{C})$.

From the homotopy theoretical point of view categories with model approximations have similar properties to those of model categories. They admit homotopy categories (localizations with respect to weak equivalences). They also can be used to construct derived functors by taking the analogs of fibrant and cofibrant replacements.

A category with weak equivalences can have several useful model approximations. We take advantage of this possibility and in each situation choose one that suits our needs. In this way we prove all the fundamental properties of the homotopy colimit and limit: Fubini Theorem (the homotopy colimit -respectively limit- commutes with itself), Thomason's theorem about diagrams indexed by Grothendieck constructions, and cofinality statements. Since the model approximations we present here consist of certain functors "indexed by spaces", the key role in all our arguments is played by the geometric nature of the indexing categories.

Received by the editor March 29, 2000, and in revised form June 27, 2000.

2000 *Mathematics Subject Classification.* Primary 55U35, 18G55; Secondary 18G10, 18F05, 55U30, 55P65.

Key words and phrases. model category, model approximation, homotopy colimit, derived functor, Grothendieck construction, Kan extension.

Part of this work has been achieved while the first author was a post-doctoral fellow at the Fields Institute, Toronto, and the Max Planck Institute, Bonn and the second at the ETH, Zürich, and the CRM, Barcelona.

The first author was supported in part by National Science Foundation grant DMS-9803766.

The second author was supported in part by Swiss National Science Foundation grant 81LA-51213.

Research partially supported by the Volkswagenstiftung Oberwolfach.

Research partially supported by Gustafsson Foundation and KTH, Stockholm.

Introduction

The purpose of this paper is to give general homotopy theoretical methods for studying diagrams. We have aimed moreover at developing tools that would provide a convenient framework for studying constructions like: push-outs and pull-backs, realizations of simplicial and cosimplicial objects, classifying spaces, orbit spaces and Borel constructions of group actions, fixed points and homotopy fixed points of groups actions, and singular chains on spaces. All these constructions are very fundamental in homotopy theory and they all are obtained by taking colimits or limits of certain diagrams with values in various categories.

One way of organizing homotopy theoretical information is by giving an appropriate model structure on the considered category. Model categories were introduced in the late sixties by D. Quillen in his foundational book [**41**]. The key roles are played by three classes of morphisms called weak equivalences, fibrations, and cofibrations, which are subject to five simple axioms (see Section 2). An important property of model categories is that one can invert the weak equivalences, so as to get the homotopy category. Model categories are also very convenient for constructing derived functors using cofibrant and fibrant replacements (non-abelian analogs of projective and injective resolutions).

This way of thinking about homotopy theory has become very popular. For example, recent advances in localization theory (see in particular [**3, 4, 5, 9, 13, 33**]) show that the category of spaces or spectra can be equipped with various model category structures, depending on what one wants to focus on. The weak equivalences for example can be chosen to be the homology equivalences for a certain homology theory. In this way our attention is placed on these properties which can be detected by the chosen homology theory.

Although model categories provide a very convenient way of doing homotopy theory, such structures are difficult to obtain. For example, for a small category I and a model category $\mathcal{M}$, quoting [**31**, page 121] "..., it seems unlikely that $Fun(I, \mathcal{M})$ has a natural model category structure." Thus to study the homotopy theory of diagrams we can not use the machinery of model categories directly. Instead our approach is to relax some of the conditions imposed on a model category so that the new structure is preserved by taking a functor category. At the same time we are not going to give up too much. We will still be able to form the localized homotopy category and construct derived functors by taking certain analogs of the cofibrant and fibrant replacements.

Our methods provide a solution to the problem that motivated us originally:

How to construct the derived functors of the colimit and limit (the homotopy colimit and limit) in any model category?

These constructions have played important roles for example in the study of classifying spaces of compact Lie groups. Started in [**43**] and continued in [**30,**

17, 36, 37, 46] several homological decompositions of classifying spaces have been found. Such a decomposition is a weak equivalence in a certain model structure (in this case it is a certain homology equivalence) between BG and the *homotopy colimit* of a diagram whose values are the classifying spaces of proper subgroups of G.

In the case of classical homotopy theory the construction of the homotopy colimit and limit has been given by A. K. Bousfield and D. Kan in [**7**] (see also [**50**]). In this case the category $Fun(I, Spaces)$ can be given *two* model structures (see [**21**, Section 2]). One where weak equivalences and *fibrations* are the objectwise weak equivalences and fibrations, and the other one where weak equivalences and *cofibrations* are the objectwise weak equivalences and cofibrations. The left derived functor of the colimit, for example, can then be obtained as follows: for a given diagram F, take its cofibrant replacement in the first model structure and compute the usual colimit. This is indeed the way homotopy push-outs have been defined for decades: before taking the colimit, replace the given push-out diagram by a weakly equivalent one, where all the objects are cofibrant and both maps are cofibrations (see [**31**, Proposition 10.6]). Similar methods were successfully applied in the category of spectra in [**49**, Section 3], and more generally in any cofibrantly generated model category, see [**34**, Theorem 14.7.1]. The same idea appears also in [**32**]. For an arbitrary model category $\mathcal{M}$, by [**31**], this approach still works when I is "very small", for example when $Fun(I, \mathcal{M})$ is the category of push-out diagrams in $\mathcal{M}$ (see Example 4.2). It fails however for G-objects in $\mathcal{M}$, where G is a finite group.

A different solution was given in the work of C. Reedy [**44**]. He introduced certain conditions on a small category I which guarantee that, for any model category $\mathcal{M}$, $Fun(I, \mathcal{M})$ can be given a model structure. This structure is good for constructing both homotopy colimits and limits. An example of such a category is given by Δ, the category of finite ordered sets.

A solution to the problem of constructing homotopy colimits and limits in any model category has been finally given in the recent work of W. Dwyer, P. Hirschhorn, and D. Kan [**18**] by using frames. The same constructions appear also in [**34**].

Before we explain our approach, we would like to give proper credit to the people that rendered the subject accessible to us and placed the landmarks in our scenery of homotopy theory of diagrams. It all started with A. K. Bousfield and D. Kan [**7**]. A systematic study of homotopy properties of diagrams has been done by W. Dwyer and D. Kan in an extensive list of papers that includes [**20**]–[**29**]. Many of their ideas have found an echo here. The work of R. Thomason including [**47, 48**] (see also [**2**]) should also be mentioned in this context. E. Dror Farjoun [**14, 15**] and A. Zabrodsky [**16**] have made important contributions to this subject as well. Let us finally mention R. M. Vogt [**50**] and the papers [**10, 11**] about homotopy coherent diagrams by J. M. Cordier and T. Porter.

Our solution goes back, as we recently noticed, to Anderson's paper [**1**, Corollary 2.12], where the idea (but not the proofs) can be found. Even though our work has been achieved independently, the fact that the first author was a student of W. Dwyer has certainly deeply influenced our way of thinking about diagrams. In fact our method could be thought of as dual to the Dwyer-Hirschhorn-Kan method. Instead of enlarging the target category, we enlarge the source. This has two advantages. First, our construction is very small (in some sense it is minimal). Second, the developed techniques are elementary and geometric.

The general scheme is as follows: as we noticed before, it seems impossible to impose directly a model structure on $Fun(I,\mathcal{M})$. Therefore we decide to "approximate" it by a larger model category. Because it is easier to deal with categories which are not as rigid as model categories, we choose to work with categories $\mathcal{C}$ where the only fixed structure is a class of weak equivalences. We define then (cf. Definition 5.1) a *left model approximation* of $\mathcal{C}$ to be a model category $\mathcal{M}$ together with a pair of adjoint functors $\mathcal{M} \rightleftarrows \mathcal{C}$ with certain properties. This pair should be thought of as an "almost" Quillen equivalence. The flexibility lies in the fact that we may vary the approximation depending on the purpose we have in mind. Even though there is no model structure on $\mathcal{C}$, having a model approximation is good enough to find an analog of a cofibrant replacement: take an object in $\mathcal{C}$, push it into $\mathcal{M}$, take its cofibrant replacement there, and finally pull it back into $\mathcal{C}$. Our main result (Theorem 11.3) can be formulated as follows:

THEOREM. *Let $\mathcal{M} \rightleftarrows \mathcal{C}$ be a left model approximation of a category $\mathcal{C}$ with a distinguished class of weak equivalences. The category of diagrams $Fun(I,\mathcal{C})$ with objectwise weak equivalences admits then a natural model approximation as well.*

The hardest part of the theorem is of course to find a model approximation for $Fun(I,\mathcal{M})$. For this purpose, we investigate the role of the *geometry* of an indexing category in the construction of the homotopy colimit. Since for an arbitrary small category it is difficult to make precise what its geometry is, we focus on the so-called simplex categories, i.e., categories associated with simplicial sets (see Definition 6.1). In this way we can take advantage of the geometry of the underlying space. However, simplex categories are big and complicated. Thus we simplify the situation by putting restrictions on diagrams indexed by them. We only consider those functors that are determined by the values they take on the non-degenerate simplices and call them bounded diagrams (see Definitions 10.1 and 17.1). The key result says then that there exists an appropriate model structure on the category $Fun^b(\mathbf{K},\mathcal{M})$ of bounded diagrams indexed by the simplex category $\mathbf{K}$ of a space K (see Theorem 13.1).

We then investigate the local properties of this model structure. It turns out that the characterization of cofibrations only depends on these local properties. In this way we can avoid checking strenuous lifting properties for a general space K and prove them only for the standard simplices $\Delta[n]$. Denote next by $\mathbf{N}(I)$ the simplex category of the nerve of a small category I (see Section 6). There is a forgetful functor $\epsilon : \mathbf{N}(I) \to I$ which induces an inclusion:

$$Fun(I,\mathcal{M}) \xrightarrow{\epsilon^*} Fun^b\big(\mathbf{N}(I),\mathcal{M}\big).$$

Together with its left adjoint, it forms the desired model approximation.

The model structure on bounded diagrams is suitable for defining a functor denoted by *ocolim* (see Definition 14.1). It is the left derived functor of the colimit restricted to the category of bounded diagrams. It is not yet the desired homotopy colimit, but it has many of its good properties. The main feature of the homotopy colimit which is not shared by *ocolim* is additivity with respect to the indexing space (see Remark 14.3). The homotopy colimit $hocolim_I$ is then constructed using *ocolim*. We show that the composite:

$$Fun(I,\mathcal{M}) \xrightarrow{\epsilon^*} Fun^b\big(\mathbf{N}(I),\mathcal{M}\big) \xrightarrow{ocolim_{\mathbf{N}(I)}} Ho(\mathcal{M})$$

is the left derived functor of $colim_I$ (see Corollary 16.4). This is equivalent to taking the ordinary colimit of the analog of the cofibrant replacement we mentioned above.

We finish by stressing the three main features of our method. First, its flexibility: for various classes of indexing categories, a different model approximation of $Fun(I, \mathcal{M})$ can be used. This idea is applied to prove the so-called Fubini theorem (the homotopy colimit commutes with itself, see Theorem 24.9), as well as Thomason's Theorem about homotopy colimits of diagrams indexed by Grothendieck constructions (see Theorem 26.8). Second, the importance of the local properties, which turn out to be essential in proving, among other things, Thomason's theorem. These local properties are a reflection of the geometric nature of our arguments. They allow the construction of a cofibrant replacement in $Fun(I, \mathcal{M})$ by doing it in $Fun^b(\mathbf{N}(I), \mathcal{M})$, as we explained above. This construction is elementary and, even when there is already a model structure on $Fun(I, \mathcal{M})$, our construction is in some sense simpler than the direct one. Third, all our constructions are easily dualizable. This duality gives, for example, a way of constructing homotopy limits and proving many of its properties.

The main features of the homotopy colimit and limit that are useful for computations are Fubini and Thomason's theorems, together with cofinality statements, see Theorem 30.5. This is the reason why we decided to devote an entire chapter to these three results. There are of course many other important properties of the homotopy colimit and limit. For example that the mapping space out of a homotopy colimit is the homotopy limit of the mapping spaces. For this however we would need to discuss mapping spaces in arbitrary model categories (not simplicial ones where the notion of mapping space is already built in) and this goes beyond the scope of this paper. We intend nevertheless to come back to this question in a short sequel.

Acknowledgments: We would like to thank Brooke Shipley and particularly Dan Christensen for their comments on a preliminary version of this paper. Our thanks also go to Jeff Smith, who pointed out to the second author that the methods we originally used to construct homotopy colimits could be applied as well to construct homotopy left Kan extensions.

Many ideas about diagrams and homotopy theory have been harvested by the first author while being a student of William G. Dwyer. Both authors would like to thank Bill Dwyer for his generosity in sharing his remarkable knowledge and imagination about homotopy theory.

We finally thank the referee for the suggested improvements in the exposition.

CHAPTER I

Model approximations and bounded diagrams

1. Notation

The symbol Δ denotes the simplicial category (cf. [**40**, section 2]), in which the objects are the ordered sets $[n] = \{n > \cdots > 0\}$, and the morphisms are weakly monotone maps of sets. The morphisms of Δ are generated by coface maps $d_i : [n-1] \to [n]$ and codegeneracy maps $s_i : [n+1] \to [n]$ for $0 \le i \le n$, subject to well-known cosimplicial identities. A simplicial set is then a functor $K : \Delta^{op} \to Sets$ where $Sets$ denotes the category of sets. One usually denotes the set $K[n]$ by K_n. A morphism between two simplicial sets is by definition a natural transformation of functors. A simplicial set K can be interpreted as a collection of sets $(K_n)_{n \ge 0}$ together with face maps $d_i : K_n \to K_{n-1}$ and degeneracy maps $s_i : K_n \to K_{n+1}$ which satisfy the simplicial identities. For a description of how to do homotopy theory in the category of simplicial sets see [**7**], [**12**], [**40**] and [**41**]. In this paper we use the symbol $Spaces$ to denote the category of simplicial sets, and by a space we always mean a simplicial set.

An element $\sigma \in K_n$ is called an n-dimensional simplex of K. It is said to be degenerate if there exists $\sigma' \in K_{n-1}$ and $0 \le i \le n-1$ such that $s_i\sigma' = \sigma$.

The standard n-simplex $\Delta[n]$ is an important example of a space. By definition, its set of k-simplices is given by $(\Delta[n])_k := \mathrm{mor}_\Delta([k],[n])$. There is a distinguished n-dimensional simplex ι in $\Delta[n]$, namely the unique non-degenerate one which comes from the identity map $[n] \to [n]$. The assignment $f \mapsto f(\iota)$ yields a bijection of sets $mor_{Spaces}(\Delta[n], K) \to K_n$. Thus we do not distinguish between maps $\Delta[n] \to K$ and n-simplices in K. If $\sigma \in K_n$ is a simplex, we use the same symbol $\sigma : \Delta[n] \to K$ to denote the corresponding map.

The simplicial subset of $\Delta[n]$ that is generated by the simplices $\{d_i\iota \mid 0 \le i \le n\}$ is denoted by $\partial\Delta[n]$ and called the boundary of $\Delta[n]$. The simplicial subset of $\partial\Delta[n]$ that is generated by the simplices $\{d_i\iota \mid i \ne k\}$ is denoted by $\Delta[n,k]$ and called a horn . There are obvious inclusions $\Delta[n,k] \subset \partial\Delta[n] \subset \Delta[n]$.

Let $\mathcal{C}$ be a category and I be a small category. By $Fun(I,\mathcal{C})$ we denote the category whose objects are functors indexed by I with values in $\mathcal{C}$, and whose morphisms are natural transformations. For any object $X \in \mathcal{C}$, there is a constant diagram $X : I \to \mathcal{C}$ with value X. This assignment defines a functor $\mathcal{C} \to Fun(I,\mathcal{C})$. Its left adjoint is called the colimit and is denoted by $colim_I : Fun(I,\mathcal{C}) \to \mathcal{C}$. Its right adjoint is called the limit and is denoted by $lim_I : Fun(I,\mathcal{C}) \to \mathcal{C}$. If this left (respectively right) adjoint exists for any small category I, we say that $\mathcal{C}$ is closed under colimits (respectively limits).

Let $F : I \to \mathcal{C}$ be a functor. The object $colim_I F$ is equipped with a natural transformation $F \to colim_I F$, from F to the constant diagram with value $colim_I F$. This natural transformation has the following universal property. For an object $X \in \mathcal{C}$, any natural transformation $F \to X$ factors uniquely as $F \to colim_I F \to X$.

For a detailed exposition on colimits and limits we refer the reader to [**31, 39**]. The colimit (respectively the limit) is a particular example of a more general left (respectively right) Kan extension. For these and other categorical constructions used in this paper see Appendix B.

Let $\mathcal{C}$ be a category and W a class of morphisms in $\mathcal{C}$. We say that W satisfies the "two out of three" property when for any composable morphisms $f : X \to Y$ and $g : Y \to Z$ in $\mathcal{C}$, if two out of f, g, and $g \circ f$ belong to W, then so does the third. A category with weak equivalences is by definition a category with a distinguished class of morphisms that contains all isomorphisms and satisfies the "two out of three" property. We use the symbol "$\xrightarrow{\sim}$" to denote a morphism in this class.

Let $\mathcal{C}$ be a category with weak equivalences. A functor $\mathcal{C} \to Ho(\mathcal{C})$ is called the localization of $\mathcal{C}$ with respect to weak equivalences if it satisfies the following universal property:

- weak equivalences in $\mathcal{C}$ are sent via $\mathcal{C} \to Ho(\mathcal{C})$ to isomorphisms (this functor is homotopy invariant);
- if $\mathcal{C} \to \mathcal{E}$ is another functor which sends weak equivalences to isomorphisms, then it can be expressed uniquely as a composite $\mathcal{C} \to Ho(\mathcal{C}) \to \mathcal{E}$ (where $\mathcal{C} \to Ho(\mathcal{C})$ is the localization).

We say that a category with weak equivalences $\mathcal{C}$ admits a localization if the functor $\mathcal{C} \to Ho(\mathcal{C})$ exists.

Let $\mathcal{C}$ be a category with weak equivalences and I be a small category. Let $\Psi : F \to G$ be a natural transformation between functors $F : I \to \mathcal{C}$ and $G : I \to \mathcal{C}$. We say that Ψ is a weak equivalence if it is an objectwise weak equivalence, i.e., if for any $i \in I$, $\Psi_i : F(i) \to G(i)$ is a weak equivalence in $\mathcal{C}$. In this way $Fun(I, \mathcal{C})$ becomes a category with weak equivalences.

2. Model categories

In this section we review classical homotopical properties of the coproduct, push-out, and the sequential colimit constructions. In order to be able to consider their *homotopical* properties, we look at these constructions in *model categories*, i.e., in categories in which one can do homotopy theory. We refer the reader to [**35, 31, 41, 42**] for the necessary definitions and theorems concerning these categories. Here we just sketch some of their properties. However the reader should keep in mind that the notion of a model category is essential in this exposition; in fact this paper is about model categories.

A model category is a category, which we usually denote by $\mathcal{M}$, together with three distinguished classes of morphisms: *weak equivalences* , *fibrations*, and *cofibrations* . This structure is subject to five axioms **MC1-MC5** (see [**31**, Section 3]). A morphism which is both a weak equivalence and a fibration (respectively a cofibration) is called an *acyclic* fibration (respectively an *acyclic* cofibration). To denote a weak equivalence, a cofibration, and a fibration we use respectively the symbols "$\xrightarrow{\sim}$","$\hookrightarrow$", and "$\twoheadrightarrow$".

Axiom **MC1** guarantees that model categories are equipped with *arbitrary* colimits and limits. In particular there is a terminal object, denoted by $*$, as well as an initial object, denoted by $\emptyset$. An object A is said to be cofibrant if the morphism $\emptyset \to A$ is a cofibration. It is said to be fibrant if the morphism $A \to *$ is a fibration. This axiom also implies the existence of products and coproducts in $\mathcal{M}$, denoted respectively by the symbols "$\prod$" and "$\coprod$".

Axiom **MC2** asserts that the class of weak equivalences satisfies the "two out of three" property. Explicitly, for two composable morphisms $f : X \to Y$ and $g : Y \to Z$, if two out of f, g, and $g \circ f$ are weak equivalences, then so is the third.

Axiom **MC3** guarantees that weak equivalences, fibrations, and cofibrations are closed under retracts. In a commutative diagram:

$$\begin{array}{ccccc} & & \overbrace{\qquad\qquad\qquad\qquad}^{id} & & \\ A & \longrightarrow & A' & \longrightarrow & A \\ {\scriptstyle f}\downarrow & & \downarrow{\scriptstyle h} & & \downarrow{\scriptstyle f} \\ B & \longrightarrow & B' & \longrightarrow & B \\ & & \underbrace{\qquad\qquad\qquad\qquad}_{id} & & \end{array}$$

if h is a weak equivalence, a fibration, or a cofibration, then so is f.

Cofibrations and fibrations are linked together through the lifting axiom **MC4**. It says that in a commutative square:

$$\begin{array}{ccc} A & \longrightarrow & X \\ {\scriptstyle i}\downarrow & & \downarrow{\scriptstyle p} \\ B & \longrightarrow & Y \end{array}$$

if either $i : A \to B$ is an acyclic cofibration and $p : X \to Y$ is a fibration, or $i : A \to B$ is a cofibration and $p : X \to Y$ is an acyclic fibration, then there is a lift, i.e., a morphism $h : B \to X$ such that the resulting diagram with five arrows commutes. In such a situation we say that $i : A \to B$ has the lifting property with respect to $p : X \to Y$. This lifting property characterizes cofibrations and fibrations. A morphism $i : A \to B$ is a cofibration if it has the lifting property with respect to all acyclic fibrations. It is an acyclic cofibration if it has the lifting property with respect to all fibrations. Dually $p : X \to Y$ is a fibration if all acyclic cofibrations have the lifting property with respect to it. It is an acyclic fibration if all cofibrations have the lifting property with respect to it.

By checking the lifting criterion one can show (see [**31**, Proposition 3.13]):

2.1. PROPOSITION. *Let the following be a push-out square in $\mathcal{M}$:*

$$\begin{array}{ccc} A & \longrightarrow & C \\ \downarrow & & \downarrow \\ B & \longrightarrow & D \end{array}$$

If $A \to B$ is an (acyclic) cofibration, then so is $C \to D$. □

Axiom **MC5** says that any morphism can be expressed as a composite of a cofibration followed by an acyclic fibration, and as a composite of an acyclic cofibration followed by a fibration. Sometimes we assume in addition that such factorizations can be chosen functorially.

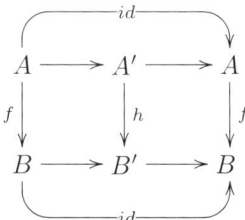

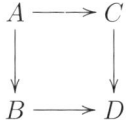

By factoring the morphism $\emptyset \to A$ into a cofibration followed by an acyclic fibration $\emptyset \hookrightarrow QA \overset{\sim}{\twoheadrightarrow} A$ we get a *cofibrant* object QA weakly equivalent to A. Such an object is called a cofibrant replacement of A. For any morphism $f : A \to B$, by the lifting axiom **MC4**, there exists $Qf : QA \to QB$ which makes the composites $QA \overset{\sim}{\twoheadrightarrow} A \xrightarrow{f} B$ and $QA \xrightarrow{Qf} QB \overset{\sim}{\twoheadrightarrow} B$ equal. Any such morphism Qf is called a cofibrant replacement of f. We say that we have chosen a cofibrant replacement Q in $\mathcal{M}$, if, for every object A, we have chosen a cofibrant replacement $QA \overset{\sim}{\twoheadrightarrow} A$ and, for every morphism $f : A \to B$, we have chosen $Qf : QA \to QB$.

Dually, by factoring the morphism $A \to *$ into an acyclic cofibration and a fibration $A \overset{\sim}{\hookrightarrow} RA \twoheadrightarrow *$ we get a *fibrant* object RA weakly equivalent to A. Such an object is called a fibrant replacement of A. For any morphism $f : A \to B$, by the lifting axiom **MC4**, there exists $Rf : RA \to RB$ which makes the composites $A \xrightarrow{f} B \overset{\sim}{\hookrightarrow} RB$ and $A \overset{\sim}{\hookrightarrow} RA \xrightarrow{Rf} RB$ equal. Any such morphism Rf is called a fibrant replacement of f. We say that we have chosen a fibrant replacement R in $\mathcal{M}$, if, for every object A, we have chosen a fibrant replacement $A \overset{\sim}{\hookrightarrow} RA$ and, for every morphism $f : A \to B$, we have chosen $Rf : RA \to RB$.

A model category is set up for defining the notion of homotopy between morphisms. For this purpose one uses so-called cylinder and mapping objects. A cylinder object of X is an object $Cyl(X)$ which fits into a factorization of the fold morphism $\nabla : X \coprod X \to X$ into a cofibration followed by a weak equivalence $X \coprod X \overset{i}{\hookrightarrow} Cyl(X) \overset{\sim}{\to} X$. A left homotopy from $f : X \to Y$ to $g : X \to Y$ is a morphism $H : Cyl(X) \to Y$ for which the following triangle commutes:

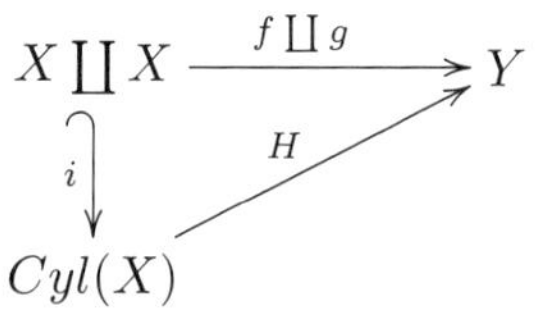

Dually, a mapping object of Y is an object $Map(Y)$ which fits into a factorization of the diagonal $\Delta : Y \to Y \times Y$ into a weak equivalence followed by a fibration $Y \overset{\sim}{\to} Map(Y) \overset{p}{\twoheadrightarrow} Y \times Y$. A right homotopy from $f : X \to Y$ to $g : X \to Y$ is a morphism $G : X \to Map(Y)$ for which the following triangle commutes:

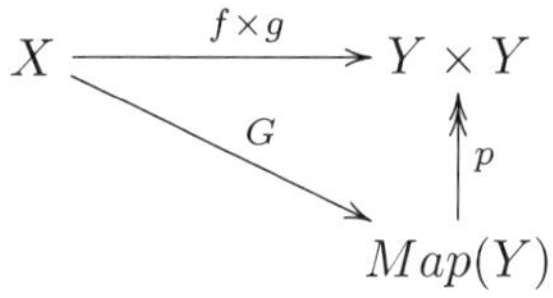

For example, given a morphism $f : X \to Y$, any two cofibrant replacements $Qf : QX \to QY$ and $Q'f : QX \to QY$ are left homotopic. Dually, any two fibrant replacements $Rf : RX \to RY$ and $R'f : RX \to RY$ of f are right homotopic.

In the case X is cofibrant and Y is fibrant both notions of right and left homotopy between morphisms $X \to Y$ coincide (see [**31**, Lemma 4.21]) and define an equivalence relation. We use the symbol "$\simeq$" to denote this relation. The homotopy category $Ho(\mathcal{M})$ is then a category having the same objects as $\mathcal{M}$, but where the morphisms from X to Y are the homotopy classes (right or left) from a *cofibrant* replacement QX of X to a *fibrant-cofibrant* replacement RQY of Y. We denote this set by $[X, Y]$. By sending an object $X \in \mathcal{M}$ to the same object $X \in Ho(\mathcal{M})$ and a morphism $f : X \to Y$ to the (right or left) homotopy class of the composite $QX \xrightarrow{Qf} QY \to RQY$, we get a functor $\mathcal{M} \to Ho(\mathcal{M})$. This functor satisfies the universal property of the localization of $\mathcal{M}$ with respect to all weak equivalences (see [**31**, Theorem 6.2]). If $f : X \to Y$ is a morphism in $\mathcal{M}$ we use the same symbol $f : X \to Y$ to denote the induced morphism in the homotopy category $Ho(\mathcal{M})$.

2.2. PROPOSITION. *Let A and B be cofibrant objects. A morphism $f : A \to B$ is a weak equivalence if and only if, for any fibrant object X and any morphism $\alpha : A \to X$, there exists a morphism $\beta : B \to X$, unique up to homotopy, such that α is homotopic to $\beta \circ f$.* □

2.3. PROPOSITION. *Let A and B be cofibrant, X fibrant, $i : A \hookrightarrow B$ a cofibration, and the following be a diagram that commutes up to homotopy:*

$$\begin{array}{ccc} A & \overset{i}{\hookrightarrow} & B \\ {\scriptstyle f}\downarrow & \swarrow {\scriptstyle g} & \\ X & & \end{array}$$

Then there exists $B \to X$, homotopic to g, which makes this diagram strictly commutative.

PROOF. The morphisms $g \circ i$ and f are homotopic, hence there exists a (right) homotopy $H : A \to Map(X)$ between them. Since X is a fibrant object, the composite $Map(X) \twoheadrightarrow X \times X \overset{p_1}{\twoheadrightarrow} X$ is an acyclic fibration. Thus there exists a lift $G : B \to Map(X)$ in the following square:

$$\begin{array}{ccc} A & \xrightarrow{H} & Map(X) \\ {\scriptstyle i}\big\downarrow & & \big\downarrow {\scriptstyle p_1}\ \sim \\ B & \xrightarrow{g} & X \end{array}$$

This lift is a homotopy from g to the composite $B \xrightarrow{G} Map(X) \xrightarrow{p_0} X$, which gives the desired morphism. □

Taking the push-out of an (acyclic) cofibration along any morphism is again an (acyclic) cofibration (see Proposition 2.1). Taking the push-out of a weak equivalence along a cofibration in general is no longer a weak equivalence. If it is so the category $\mathcal{M}$ is called left proper (see [**6**, Definition 1.2]). Nevertheless in the case the involved objects are cofibrant, we have:

2.4. PROPOSITION. *Let $A \hookrightarrow B$ be a cofibration, $f : A \xrightarrow{\sim} C$ be a weak equivalence, and the following be a push-out square:*

$$\begin{array}{ccc} A & \overset{i}{\hookrightarrow} & B \\ {\scriptstyle f}\big\downarrow \sim & & \big\downarrow {\scriptstyle g} \\ C & \overset{j}{\hookrightarrow} & D \end{array}$$

If A, B, and C are cofibrant, then $g : B \to D$ is a weak equivalence.

PROOF. Since C is cofibrant and j is a cofibration, D is cofibrant. Hence, by Proposition 2.2, we have to show that for a fibrant object X and a morphism $\alpha : B \to X$, there exists $\beta : D \to X$, unique up to homotopy, such that $\beta \circ g \simeq \alpha$. Since f is a weak equivalence, there is a $\gamma : C \to X$ for which $i \circ \alpha \simeq \gamma \circ f$. By assumption i is a cofibration, thus according to Proposition 2.3 there exists $\alpha' : B \to X$, homotopic to α, such that $\gamma \circ f = \alpha' \circ i$. In this way we obtain a morphism $\beta : D \to X$, from the push-out D, with the desired property.

The uniqueness can be checked in a similar way by replacing X, in the previous argument, with $Map(X)$. □

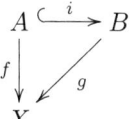

Propositions 2.2, 2.3, and 2.4 can be used to study homotopy invariance of the coproduct, push-out, and the sequential colimit constructions. For a transfinite

telescope diagram of the form $A_0 \to A_1 \to A_2 \to \cdots$, we require as usual that the value at a limit ordinal γ be the canonical one, i.e., $A_\gamma = colim_{\beta<\gamma} A_\beta$.

2.5. PROPOSITION.

1. *If for $i \in I$, $f_i : X_i \to Y_i$ is a weak equivalence between cofibrant objects, then so is the coproduct $\coprod_{i\in I} f_i : \coprod_{i\in I} A_i \to \coprod_{i\in I} B_i$.*
2. *Consider the following natural transformation between push-out diagrams:*

$$\begin{array}{ccccccc} A & = & colim\ (& A_0 & \longleftarrow & A_1 \hookrightarrow & A_2\ ) \\ \downarrow f & & & f_0 \downarrow \sim & & \sim \downarrow f_1 & \sim \downarrow f_2 \\ B & = & colim\ (& B_0 & \longleftarrow & B_1 \longrightarrow & B_2\ ) \end{array}$$

where, for $k = 0,1,2$, f_k is a weak equivalence and A_k, B_k are cofibrant. If $A_1 \hookrightarrow A_2$ and either $B_1 \to B_2$ or $B_1 \to B_0$ are cofibrations, then $f : A \to B$ is a weak equivalence.

3. *Consider the following natural transformation between (possibly transfinite) telescope diagrams:*

$$\begin{array}{ccccccc} A & = & colim\ (& A_0 \overset{i_0}{\hookrightarrow} & A_1 \overset{i_1}{\hookrightarrow} & A_2 \overset{i_2}{\hookrightarrow} \cdots\ ) \\ \downarrow f & & & f_0 \downarrow \sim & \sim \downarrow f_1 & \sim \downarrow f_2 \\ B & = & colim\ (& B_0 \overset{j_0}{\hookrightarrow} & B_1 \overset{j_1}{\hookrightarrow} & B_2 \overset{j_2}{\hookrightarrow} \cdots\ ) \end{array}$$

where, for $k \geq 0$, f_k is a weak equivalence, A_k and B_k are cofibrant, and i_k, j_k are cofibrations. Then $f : A \to B$ is a weak equivalence.

We show only 3 since 1 and 2 can be proved using similar methods.

PROOF OF 3. The colimit A is equipped with morphisms $\zeta_k : A_k \to A$ for all $k \geq 0$. By checking the lifting property, it is easy to see that $\zeta_0 : A_0 \to A$ is a cofibration and hence A is cofibrant. We can then apply Proposition 2.2 to prove that f is a weak equivalence. Let X be fibrant and $\alpha : A \to X$ be a morphism. Since f_0 is a weak equivalence, there is $\beta_0 : B_0 \to X$ such that $\beta_0 \circ f_0 \simeq \alpha \circ \zeta_0$. Using Proposition 2.3 we can modify α, up to homotopy, to get $\alpha_0 : A \to X$ such that $\beta_0 \circ f_0 = \alpha_0 \circ \zeta_0$.

In the next step, since f_1 is a weak equivalence, we can find $\beta_1' : B_1 \to X$ for which $\beta_1' \circ f_1 \simeq \alpha_0 \circ \zeta_1$. By precomposing with i_0 we see that $\beta_0 \circ f_0$ is homotopic to $\beta_1' \circ j_0 \circ f_0$. Therefore $\beta_0 \simeq \beta_1' \circ j_0$. Hence we can replace β_1' by a homotopic morphism $\beta_1 : B_1 \to X$ which gives a strict equality $\beta_0 = \beta_1 \circ j_0$. We can again modify α_0, up to homotopy, to get $\alpha_1 : A \to X$ such that $\beta_1 \circ f_1 = \alpha_1 \circ \zeta_1$.

Continuing this process inductively we get a family of strictly compatible morphisms $\beta_k : B_k \to X$ inducing $\beta : B \to X$. To construct a homotopy between $\beta \circ f$ and α and to show the homotopical uniqueness of such β one can use a similar argument replacing X with $Map(X)$. □

Under some circumstances the coproduct, push-out, and the sequential colimit constructions preserve also cofibrations and acyclic cofibrations.

2.6. PROPOSITION.

1. *If for all $i \in I$, $f_i : A_i \to B_i$ is an (acyclic) cofibration, then so is the coproduct $\coprod_{i\in I} f_i : \coprod_{i\in I} A_i \to \coprod_{i\in I} B_i$.*

2. *Consider the following natural transformation between push-out diagrams :*

$$\begin{array}{ccccccc} A & = & colim & (A_0 & \longleftarrow & A_1 & \longrightarrow & A_2) \\ {\scriptstyle f}\downarrow & & & {\scriptstyle f_0}\downarrow & & \downarrow{\scriptstyle f_1} & & \downarrow{\scriptstyle f_2} \\ B & = & colim & (B_0 & \longleftarrow & B_1 & \longrightarrow & B_2) \end{array}$$

Let $M = colim(B_1 \xleftarrow{f_1} A_1 \to A_2)$ *and* $g : M \to B_2$ *be induced by the commutativity of the above diagram. Then:*
- *If* f_0 *and* g *are cofibrations, then so is* f.
- *If* f_0, f_1 *are acyclic cofibrations,* f_2 *is a weak equivalence and* g *is a cofibration, then* g *and* f *are acyclic cofibrations.*

3. *Consider the following natural transformation between (possibly transfinite) telescope diagrams :*

$$\begin{array}{ccccccccc} A & = & colim & (A_0 & \longrightarrow & A_1 & \longrightarrow & A_2 & \longrightarrow \cdots) \\ {\scriptstyle f}\downarrow & & & {\scriptstyle f_0}\downarrow & & \downarrow{\scriptstyle f_1} & & \downarrow{\scriptstyle f_2} & \\ B & = & colim & (B_0 & \longrightarrow & B_1 & \longrightarrow & B_2 & \longrightarrow \cdots) \end{array}$$

If f_0 *and, for all* i, $colim(B_i \leftarrow A_i \to A_{i+1}) \to B_{i+1}$ *are (acyclic) cofibrations, then so is* f.

Since the proofs are analogous we show only the second part of 2.

PROOF OF THE SECOND PART OF 2. The morphism $f_2 : A_2 \to B_2$ factors as a composite $A_2 \to M \xrightarrow{g} B_2$. By assumption f_1 is an acyclic cofibration and hence so is $A_2 \to M$. Therefore, since f_2 is a weak equivalence, g is an *acyclic* cofibration.

To prove that f is an acyclic cofibration we have to show that it has the lifting property with respect to all fibrations, i.e., for any fibration $X \twoheadrightarrow Y$ and any commutative square:

$$\begin{array}{ccc} A & \longrightarrow & X \\ {\scriptstyle f}\downarrow & & \downarrow \\ B & \longrightarrow & Y \end{array} \qquad \text{we need to construct a lift } h : B \to X.$$

For this purpose consider the following commutative diagram:

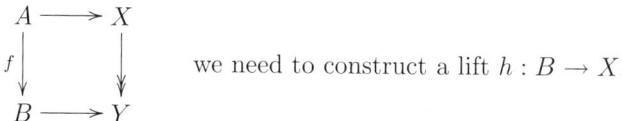

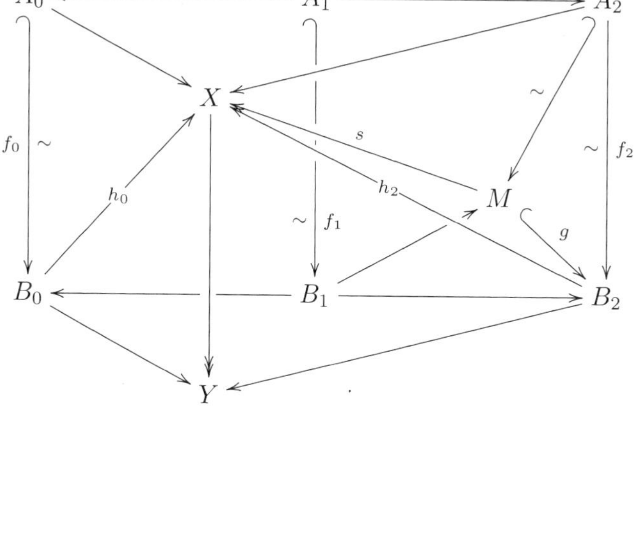

where:

- the morphism $h_0 : B_0 \to X$ is constructed by the lifting property of the acyclic cofibration $f_0 : A_0 \overset{\sim}{\hookrightarrow} B_0$ with respect to the fibration $X \twoheadrightarrow Y$;
- the morphism $s : M \to X$ is then induced by the universal property of the colimit $M = colim(B_1 \leftarrow A_1 \to A_2)$;
- finally, $h_2 : B_2 \to X$ is constructed by the lifting property of the acyclic cofibration $g : M \overset{\sim}{\hookrightarrow} B_2$ with respect to the fibration $X \twoheadrightarrow Y$.

The desired lift $h : B \to X$ can be now constructed using the morphisms h_0 and h_2. □

3. Left derived functors

In this section we recall the notion of a left derived functor (see [**31**, Definition 9.1] and [**41**, Definition 4.1]). Let $\mathcal{D}$ be a category with weak equivalences (see Section 1). In the case $\mathcal{D}$ is a model category we focus our attention only on the class of weak equivalences.

3.1. Definition. We say that a functor $\mathcal{H} : \mathcal{D} \to \mathcal{E}$ is *homotopy invariant* if, for any weak equivalence f in $\mathcal{D}$, $\mathcal{H}(f)$ is an isomorphism in $\mathcal{E}$.

If the category $\mathcal{D}$ admits a localization functor with respect to weak equivalences $\mathcal{D} \to Ho(\mathcal{D})$, then, by the universal property, $\mathcal{H} : \mathcal{D} \to \mathcal{E}$ is homotopy invariant if and only if it can be expressed as a composite $\mathcal{D} \to Ho(\mathcal{D}) \to \mathcal{E}$.

Functors which are not homotopy invariant can be often approximated by ones that are so, the so-called derived functors, which are defined by a universal property:

3.2. Definition. A functor $L(\mathcal{H}) : \mathcal{D} \to \mathcal{E}$ together with a natural transformation $L(\mathcal{H}) \to \mathcal{H}$ is called the *left derived functor* of $\mathcal{H} : \mathcal{D} \to \mathcal{E}$ if:

- $L(\mathcal{H})$ is homotopy invariant;
- if $\mathcal{G} : \mathcal{D} \to \mathcal{E}$ is a homotopy invariant functor, then any natural transformation $\mathcal{G} \to \mathcal{H}$ factors uniquely as a composite $\mathcal{G} \to L(\mathcal{H}) \to \mathcal{H}$.

Let $\mathcal{C}$ be a category with weak equivalences that admits a localization $Ho(\mathcal{C})$ (for example a model category) and $\mathcal{H} : \mathcal{D} \to \mathcal{C}$ be a functor. The left derived functor of the composite $\mathcal{D} \xrightarrow{\mathcal{H}} \mathcal{C} \to Ho(\mathcal{C})$ is called the *total* left derived functor of $\mathcal{H}$ (see [**31**, Definition 9.5]) and is also denoted by the symbol $L(\mathcal{H})$.

The essential data needed to define the total left derived functor of $\mathcal{H} : \mathcal{D} \to \mathcal{C}$ is the choice of weak equivalences in $\mathcal{D}$. For its construction however, it is very helpful to have some additional structure on $\mathcal{D}$. For example model categories have been invented for this purpose. In this section we outline the standard way of building left derived functors in the case $\mathcal{D}$ is a model category by using cofibrant replacements. In Section 5 we generalize this to categories with model approximations.

3.3. Definition. Let $\mathcal{M}$ be a model category and $\mathcal{C}$ be a category with weak equivalences. A functor $\mathcal{H} : \mathcal{M} \to \mathcal{C}$ is called *homotopy meaningful on cofibrant objects* if, for any weak equivalence $f : X \to Y$ in $\mathcal{M}$ between cofibrant objects X and Y, $\mathcal{H}(f)$ is a weak equivalence in $\mathcal{C}$.

A convenient test for verifying that a functor is homotopy meaningful on cofibrant objects is given by K. Brown's lemma (see [**8**] and [**31**, Lemma 9.9]).

3.4. PROPOSITION (K. Brown). *Let $\mathcal{M}$ be a model category and $\mathcal{C}$ be a category with weak equivalences. Then a functor $\mathcal{H} : \mathcal{M} \to \mathcal{C}$ is homotopy meaningful on cofibrant objects if and only if for any acyclic cofibration $f : X \overset{\sim}{\hookrightarrow} Y$ in $\mathcal{M}$, where X is cofibrant, $\mathcal{H}(f)$ is a weak equivalence in $\mathcal{C}$*

PROOF. We only have to check that the condition given in the proposition is sufficient. Let $f : X \overset{\sim}{\to} Y$ be a weak equivalence between cofibrant objects in $\mathcal{M}$. Let $X \coprod Y \hookrightarrow QY \overset{\sim}{\twoheadrightarrow} Y$ be the factorization of $f \coprod id$ into a cofibration followed by an acyclic fibration. Consider the following commutative diagrams respectively in $\mathcal{M}$ and $\mathcal{C}$:

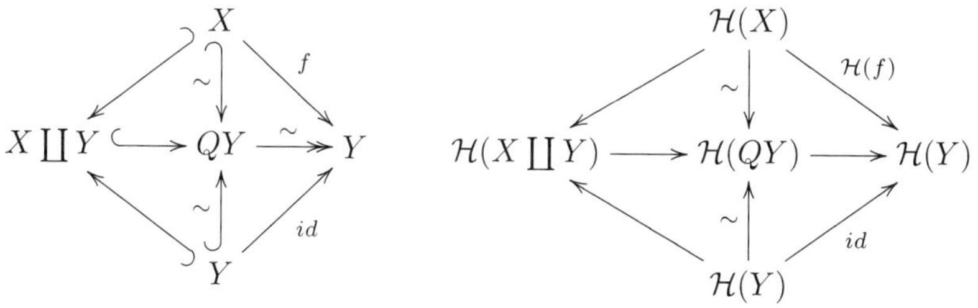

Since $X \overset{\sim}{\hookrightarrow} QY$ and $Y \overset{\sim}{\hookrightarrow} QY$ are acyclic cofibrations in $\mathcal{M}$, the morphisms $\mathcal{H}(X) \to \mathcal{H}(QY)$ and $\mathcal{H}(Y) \to \mathcal{H}(QY)$ are weak equivalences in $\mathcal{C}$. Thus the "two out of three" property implies that $\mathcal{H}(QY) \to \mathcal{H}(Y)$ is also a weak equivalence, and hence, by the same argument, so is $\mathcal{H}(f) : \mathcal{H}(X) \to \mathcal{H}(Y)$. □

3.5. PROPOSITION. *Let $\mathcal{C}$ be a category with weak equivalences that admits a localization $Ho(\mathcal{C})$. Let $\mathcal{M}$ be a model category. If $\mathcal{H} : \mathcal{M} \to \mathcal{C}$ is homotopy meaningful on cofibrant objects, then its total left derived functor $L(\mathcal{H})$ exists. It can be constructed by choosing a cofibrant replacement Q in $\mathcal{M}$ and assigning to $X \in \mathcal{M}$ the object $L(\mathcal{H})(X) := \mathcal{H}(QX) \in Ho(\mathcal{C})$. The natural transformation $L(\mathcal{H}) \to \mathcal{H}$ is induced by the morphisms $\mathcal{H}(QX \overset{\sim}{\twoheadrightarrow} X)$.*

3.6. LEMMA. *Let $\mathcal{C}$ and $\mathcal{M}$ be as in Proposition 3.5 and $\mathcal{H} : \mathcal{M} \to \mathcal{C}$ be homotopy meaningful on cofibrant objects. If $f : X \to Y$ and $g : X \to Y$ are left homotopic in $\mathcal{M}$, and X is cofibrant, then $\mathcal{H}(f) = \mathcal{H}(g)$ in $Ho(\mathcal{C})$.*

PROOF. Since f and g are left homotopic, we can form a commutative diagram in $\mathcal{M}$:

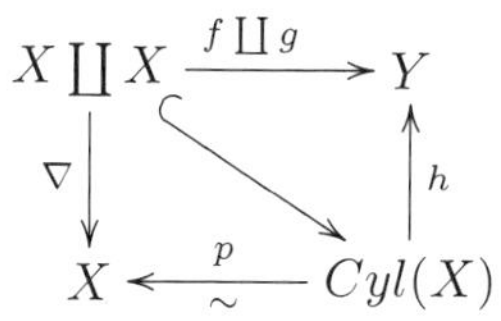

By applying $\mathcal{H}$ we get the following commutative diagram in $Ho(\mathcal{C})$:

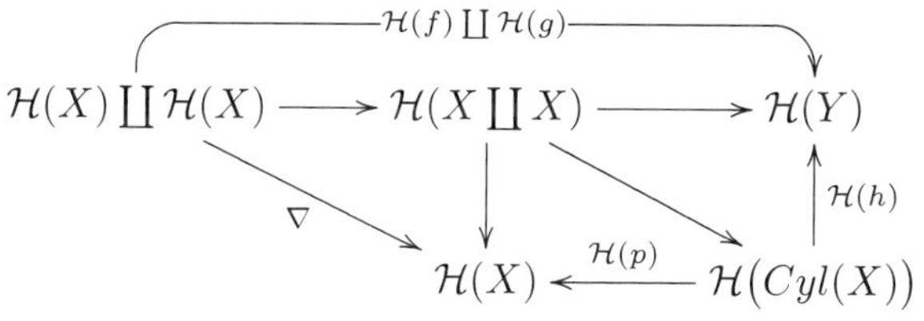

Since X is cofibrant, then so is $Cyl(X)$. As $\mathcal{H}$ is homotopy meaningful on cofibrant objects, $\mathcal{H}(p)$ is an isomorphism in $Ho(\mathcal{C})$. It follows that $\mathcal{H}(f)$ and $\mathcal{H}(g)$ are both equal to the composite $\mathcal{H}(X) \xrightarrow{\mathcal{H}(p)^{-1}} \mathcal{H}(Cyl(X)) \xrightarrow{\mathcal{H}(h)} \mathcal{H}(Y)$. □

PROOF OF PROPOSITION 3.5. Let $f : X \to Y$ and $g : Y \to Z$ be morphisms in $\mathcal{M}$. Since:

$$QX \xrightarrow{Q(g\circ f)} QZ \qquad\qquad QX \xrightarrow{Qf} QY \xrightarrow{Qg} QZ$$

are cofibrant replacements of the same morphism $g \circ f$, they are left homotopic. Thus by Lemma 3.6, the assignment $\mathcal{M} \ni X \mapsto \mathcal{H}(QX) \in Ho(\mathcal{C})$ is a well defined functor.

We have to show that the natural transformation $\mathcal{H}(QX \overset{\sim}{\twoheadrightarrow} X)$ satisfies the appropriate universal property. Let $\mathcal{G} : \mathcal{M} \to Ho(\mathcal{C})$ be a homotopy invariant functor and $\mathcal{G} \to \mathcal{H}$ be a natural transformation. Since $\mathcal{G}$ is homotopy invariant, $\mathcal{G}(QX) \to \mathcal{G}(X)$ is an isomorphism. Therefore, for any object X in $\mathcal{M}$, there exists a unique morphism $\mathcal{G}(X) \to \mathcal{H}(QX)$ for which the following diagram commutes in $Ho(\mathcal{C})$:

$$\begin{array}{ccc}
\mathcal{G}(QX) & \longrightarrow & \mathcal{H}(QX) \\
{\scriptstyle\cong}\downarrow & \nearrow & \downarrow \\
\mathcal{G}(X) & \longrightarrow & \mathcal{H}(X)
\end{array}$$

These morphisms form the desired unique natural transformation $\mathcal{G} \to \mathcal{H}(Q-)$. □

4. Left derived functors of colimits and left Kan extensions

Let I be a small category and $\mathcal{C}$ a category with weak equivalences. Let us consider the category of functors $Fun(I, \mathcal{C})$. Recall that a natural transformation $\Psi : F \to G$ is called a *weak equivalence* in $Fun(I, \mathcal{C})$ if for all $i \in I$, $\Psi_i : F(i) \to G(i)$ is a weak equivalence in $\mathcal{C}$.

Let $f : I \to J$ be a functor and $\mathcal{C}$ be closed under colimits. In general, neither the colimit construction $colim_I : Fun(I, \mathcal{C}) \to \mathcal{C}$ nor the left Kan extension $f^k : Fun(I, \mathcal{C}) \to Fun(J, \mathcal{C})$ (see Section 36) are homotopy meaningful: a weak equivalence between diagrams usually does not induce a weak equivalence on colimits or on left Kan extensions. Our key objective in this work is to construct their homotopy meaningful approximations.

4.1. DEFINITION. Let $\mathcal{C}$ be a category with weak equivalences closed under colimits and $f : I \to J$ be a functor of small categories.

- Assume that $\mathcal{C}$ admits a localization $\mathcal{C} \to Ho(\mathcal{C})$. The total left derived functor of $colim_I : Fun(I, \mathcal{C}) \to \mathcal{C}$ is called the *homotopy colimit* and is denoted by $hocolim_I : Fun(I, \mathcal{C}) \to Ho(\mathcal{C})$.
- Assume that $Fun(J, \mathcal{C})$ admits a localization. The total left derived functor of $f^k : Fun(I, \mathcal{C}) \to Fun(J, \mathcal{C})$ is called the *homotopy left Kan extension* along f.

Let $\mathcal{M}$ be a model category. What is a general strategy for constructing the total left derived functor of $colim_I : Fun(I, \mathcal{M}) \to \mathcal{M}$? According to Proposition 3.5 it suffices to put a model category structure on $Fun(I, \mathcal{M})$ for which the colimit functor is homotopy invariant on cofibrant objects.

4.2. EXAMPLE. Let I be the push-out category, i.e., the category given by the graph $(0 \leftarrow 1 \rightarrow 2)$. Let $\Psi : F \rightarrow G$ be a morphism in $Fun(I, \mathcal{M})$ given by the commutative diagram:

$$\begin{array}{ccccccc} F & = & (F(0) & \longleftarrow & F(1) & \longrightarrow & F(2)) \\ \downarrow \Psi & & \downarrow \Psi_0 & & \downarrow \Psi_1 & & \downarrow \Psi_2 \\ G & = & (G(0) & \longleftarrow & G(1) & \longrightarrow & G(2)) \end{array}$$

- We call Ψ a weak equivalence if it is an objectwise weak equivalence, i.e., if Ψ_i is a weak equivalence in $\mathcal{M}$ for $0 \leq i \leq 2$.
- We call Ψ a fibration if it is an objectwise fibration, i.e., if Ψ_i is a fibration in $\mathcal{M}$ for $0 \leq i \leq 2$.
- We call Ψ a cofibration if the morphisms $\Psi_1 : F(1) \rightarrow G(1)$ and, for $i \neq 1$, $colim\big(F(i) \leftarrow F(1) \xrightarrow{\Psi_1} G(1)\big) \rightarrow G(i)$ are cofibrations in $\mathcal{M}$.

It is not difficult to check that $Fun(I, \mathcal{M})$, with the above choice of weak equivalences, fibrations, and cofibrations, satisfies the axioms of a model category (see [**31**, Section 10] for a more detailed discussion of push-out diagrams). With this model structure a diagram $F(0) \leftarrow F(1) \rightarrow F(2)$ is cofibrant if the object $F(1)$ is cofibrant and the morphisms $F(1) \rightarrow F(0)$, $F(1) \rightarrow F(2)$ are cofibrations in $\mathcal{M}$.

Proposition 2.5 (2) implies that the functor $colim_I : Fun(I, \mathcal{M}) \rightarrow \mathcal{M}$ is homotopy meaningful on cofibrant objects. Therefore, according to Proposition 3.5, its total left derived functor $hocolim_I$ exists. It can be constructed by taking the colimit of a cofibrant replacement. Moreover, Proposition 2.5 (2) implies that if $F = \big(F(0) \leftarrow F(1) \rightarrow F(2)\big)$ is a push-out diagram of cofibrant objects where either $F(1) \rightarrow F(0)$ or $F(1) \rightarrow F(2)$ is a cofibration, then the natural morphism $hocolim_I(F) \rightarrow colim_I(F)$ is a weak equivalence.

4.3. EXAMPLE. Let I be the telescope category, i.e., the category given by the graph $(0 \rightarrow 1 \rightarrow 2 \rightarrow \cdots)$. Let $\Psi : F \rightarrow G$ be a morphism in $Fun(I, \mathcal{M})$ given by the commutative diagram:

$$\begin{array}{ccccccccc} F & = & (F(0) & \longrightarrow & F(1) & \longrightarrow & F(2) & \longrightarrow & \cdots) \\ \downarrow \Psi & & \downarrow \Psi_0 & & \downarrow \Psi_1 & & \downarrow \Psi_2 & & \\ G & = & (G(0) & \longrightarrow & G(1) & \longrightarrow & G(2) & \longrightarrow & \cdots) \end{array}$$

- We call Ψ a weak equivalence if it is an objectwise weak equivalence, i.e., if Ψ_i is a weak equivalence in $\mathcal{M}$ for $i \geq 0$.
- We call Ψ a fibration if it is an objectwise fibration, i.e., if Ψ_i is a fibration in $\mathcal{M}$ for $i \geq 0$.
- We call Ψ a cofibration if the morphisms $\Psi_0 : F(0) \rightarrow G(0)$ and, for $i \geq 0$, $colim\big(G(i) \xleftarrow{\Psi_i} F(i) \rightarrow F(i+1)\big) \rightarrow G(i+1)$ are cofibrations in $\mathcal{M}$.

It is not difficult to check that $Fun(I, \mathcal{M})$, with the above choice of weak equivalences, fibrations, and cofibrations, satisfies the axioms of a model category. With this model structure a diagram $\big(F(0) \rightarrow F(1) \rightarrow F(2) \rightarrow \cdots\big)$ is cofibrant if the object $F(0)$ is cofibrant and, for $i \geq 0$, the morphisms $F(i) \rightarrow F(i+1)$ are cofibrations.

Proposition 2.5 (3) implies that the functor $colim_I : Fun(I, \mathcal{M}) \rightarrow \mathcal{M}$ is homotopy invariant on cofibrant objects. Therefore, according to Proposition 3.5, its total left derived functor $hocolim_I$ exists. It can be constructed by taking the colimit of a cofibrant replacement.

Although model categories provide a very useful framework for constructing derived functors, model category structures themselves are difficult to obtain. For example, for general I and $\mathcal{M}$, we do not know how to put any natural model structure on $Fun(I, \mathcal{M})$, in particular one for which $colim_I$ is homotopy meaningful on cofibrant objects. Even if we are in special circumstances where $Fun(I, \mathcal{M})$ can be given such a model structure (for example when $\mathcal{M}$ is cofibrantly generated [**35**, Theorem 14.7.1]), cofibrations are usually very complicated. In such cases the construction of a cofibrant replacement of a given diagram is very involved. Thus instead of imposing a model structure directly on $Fun(I, \mathcal{M})$, we are going to approximate it by a model category. Using this approximation we can find a candidate for a "cofibrant replacement" in $Fun(I, \mathcal{C})$ (see Remark 5.10). We can then construct the total left derived functor of $colim_I$ as in Proposition 5.9. This will be achieved in Section 16. The remaining sections in this chapter are devoted to the set up of the necessary tools.

5. Model approximations

In this section we introduce the fundamental concept of this paper: that of a left model approximation. The aim is to relax some of the requirements imposed on a model category, so that the new structure would be preserved by taking a functor category.

5.1. Definition. Let $\mathcal{D}$ be a category with weak equivalences. A *left model approximation* of $\mathcal{D}$ is a model category $\mathcal{M}$ together with a pair of adjoint functors:

$$\mathcal{M} \underset{r}{\overset{l}{\rightleftarrows}} \mathcal{D}$$

where:

1. the functor l is left adjoint to r;
2. the functor r is homotopy meaningful, i.e., if f is a weak equivalence in $\mathcal{D}$, then rf is a weak equivalence in $\mathcal{M}$;
3. the functor l is homotopy meaningful on cofibrant objects;
4. for any object A in $\mathcal{D}$ and any cofibrant object X in $\mathcal{M}$, if a morphism $X \to rA$ is a weak equivalence in $\mathcal{M}$, then so is its adjoint $lX \to A$ in $\mathcal{D}$.

Observe that a model approximation is "almost" a Quillen equivalence (see [**41**, I.4.5]), where we are allowed to use only the existence of weak equivalences in $\mathcal{D}$, and not the entire model structure. In particular in the above definition we do not assume that $\mathcal{D}$ is closed under colimits.

5.2. Example. When $\mathcal{D}$ is already a model category, the identity functors $id : \mathcal{D} \rightleftarrows \mathcal{D} : id$ form a model approximation of $\mathcal{D}$. Thus the notion of a left model approximation generalizes that of a model category.

5.3. Example. The category *Spaces* together with the realization and the singular functor is a left model approximation of the category of CW-complexes. The geometric realization is left adjoint to the singular functor. In this example the approximated category is not closed under colimits.

The left model approximation of $\mathcal{D}$ is not a well-defined object. It does not have to be unique. It is actually very convenient to work with several model approximations at the same time, depending on the applications one has in mind. In Theorem 24.9 for example, we compare the results of the same computation done in two different model approximations.

5.4. REMARK. One of the key objective of this exposition is to show that from the homotopy theoretical point of view being a model category or having a model approximation does not make much difference. One can prove that in both cases we can:

- form the localized homotopy category (see Proposition 5.5);
- construct suspensions and general homotopy colimits (see Corollary 16.2);
- form Puppe's sequences;
- construct mapping spaces;
- define the notion of cofibrant replacement and thus build left derived functors (see Proposition 5.9).

In addition we show that categories with model approximations are naturally closed under taking functor categories (see Theorem 15.1).

5.5. PROPOSITION. *Let $l : \mathcal{M} \rightleftarrows \mathcal{D} : r$ be a left model approximation of $\mathcal{D}$. Then the localization of $\mathcal{D}$ with respect to weak equivalences exists. The homotopy category $Ho(\mathcal{D})$ can be constructed as follows: objects of $Ho(\mathcal{D})$ coincide with objects of $\mathcal{D}$ and $mor_{Ho(\mathcal{D})}(X, Y) := mor_{Ho(\mathcal{M})}(rX, rY)$.*

5.6. LEMMA. *Let $l : \mathcal{M} \rightleftarrows \mathcal{D} : r$ be a left model approximation of $\mathcal{D}$. A morphism $f : A \to B$ in $\mathcal{D}$ is a weak equivalence if and only if rf is a weak equivalence in $\mathcal{M}$.*

PROOF. Let us assume that rf is a weak equivalence in $\mathcal{M}$. By taking cofibrant replacements we get a commutative square where all morphisms are weak equivalences:

$$\begin{array}{ccc} QrA & \xrightarrow{Qrf} & QrB \\ \downarrow & & \downarrow \\ rA & \xrightarrow{rf} & rB \end{array}$$

Since QrA is cofibrant, the morphisms $lQrA \to A$ and $lQrA \to B$ are weak equivalences as their adjoints are so. Commutativity of the triangle:

$$\begin{array}{ccc} lQrA & & \\ \sim\downarrow & \searrow\sim & \\ A & \xrightarrow{f} & B \end{array}$$

and the "two out of three" property prove that f is a weak equivalence as well. □

PROOF OF PROPOSITION 5.5. . Let $\alpha : \mathcal{D} \to \mathcal{E}$ be a homotopy invariant functor. We are going to prove that there exists a unique functor $\beta : Ho(\mathcal{D}) \to \mathcal{E}$ for which the composite $\mathcal{D} \to Ho(\mathcal{D}) \xrightarrow{\beta} \mathcal{E}$ equals α. On objects we have no choice, we define $\beta(A) := \alpha(A)$.

Let A and B be objects in $\mathcal{D}$. Since $mor_{Ho(\mathcal{D})}(A, B) = [rA, rB]$, a morphism $[f] : A \to B$ in $Ho(\mathcal{D})$ is induced by a morphism $f : QrA \to RQrB$ in $\mathcal{M}$, where

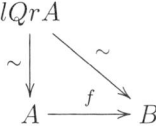

Q and R are appropriate cofibrant and fibrant replacements (see Section 2 for the definition of a morphism in the homotopy category). Consider the following sequence of morphisms in $\mathcal{D}$:

$$A \leftarrow lQrA \xrightarrow{lf} lRQrB \xleftarrow{i} lQrB \rightarrow B$$

Observe that $lQrA \to A$ and $lQrB \to B$ are weak equivalences since their adjoints are so. The morphism $lRQrB \leftarrow lQrB$ is also a weak equivalence as l is homotopy invariant on cofibrant objects. We define $\beta([f])$ to be the unique morphism which makes the following diagram commute in $\mathcal{E}$:

$$\begin{array}{ccccc}
\alpha(lQrA) & \longrightarrow & \alpha(lRQrB) & \xleftarrow{\cong} & \alpha(lQrB) \\
\downarrow{\cong} & \nearrow & & \searrow & \downarrow{\cong} \\
\alpha(A) & & \xrightarrow{\qquad\beta([f])\qquad} & & \alpha(B)
\end{array}$$

Since α converts weak equivalences into isomorphisms, $\beta([f])$ does exist. One can finally check that this process defines the desired functor $\beta : Ho(\mathcal{D}) \to \mathcal{E}$. □

Lemma 5.6 also implies:

5.7. COROLLARY. *Let $l : \mathcal{M} \rightleftarrows \mathcal{C} : r$ be a left model approximation. The class of weak equivalences in $\mathcal{C}$ is closed under retracts.* □

In a similar way as in the case of model categories, model approximations can also be used to construct left derived functors.

5.8. DEFINITION. Let $\mathcal{C}$ be a category with weak equivalences. We say that a left model approximation $l : \mathcal{M} \rightleftarrows \mathcal{D} : r$ is *good* for a functor $\mathcal{F} : \mathcal{D} \to \mathcal{C}$ if the composite $\mathcal{F} \circ l : \mathcal{M} \to \mathcal{C}$ is homotopy meaningful on cofibrant objects.

5.9. PROPOSITION. *Let $\mathcal{C}$ be a category with weak equivalences that admits a localization $\mathcal{C} \to Ho(\mathcal{C})$. Let $l : \mathcal{M} \rightleftarrows \mathcal{D} : r$ be a left model approximation which is good for $\mathcal{F} : \mathcal{D} \to \mathcal{C}$. Then the total left derived functor of $\mathcal{F}$ exists. It can be constructed by taking the composite $\mathcal{D} \xrightarrow{r} \mathcal{M} \xrightarrow{L(\mathcal{F}\circ l)} Ho(\mathcal{C})$.*

PROOF. Let us denote by $\mathcal{H} : \mathcal{M} \to \mathcal{C}$ the composite $\mathcal{F} \circ l$. Since this functor is homotopy meaningful on cofibrant objects, according to Proposition 3.5, its total left derived functor exists. Moreover, it can be constructed by choosing a cofibrant replacement Q in $\mathcal{M}$ and taking $L(\mathcal{H})(X) = \mathcal{H}(QX)$.

For any $A \in \mathcal{D}$, define:

- QA to be $lQrA$,
- $QA = lQrA \to A$ to be the adjoint of $QrA \overset{\sim}{\twoheadrightarrow} rA$.

Observe that the assignment $\mathcal{D} \ni A \mapsto \mathcal{F}(QA) \in Ho(\mathcal{C})$ is a well defined functor as it coincides with $L(\mathcal{H}) \circ r$.

We are going to show that $\mathcal{F}(Q-) : \mathcal{D} \to Ho(\mathcal{C})$, together with the natural transformation induced by $\mathcal{F}(QA \to A)$, is the total left derived functor of $\mathcal{F}$. Let $\mathcal{G} : \mathcal{D} \to Ho(\mathcal{C})$ be homotopy invariant and $\mathcal{G} \to \mathcal{F}$ be a natural transformation. For any $A \in \mathcal{D}$, define $\mathcal{G}(A) \to \mathcal{F}(QA)$ to be the unique morphism that fits into the following commutative diagram:

$$\begin{array}{ccc} \mathcal{G}(QA) & \longrightarrow & \mathcal{F}(QA) \\ \downarrow & \nearrow & \downarrow \\ \mathcal{G}(A) & \longrightarrow & \mathcal{F}(A) \end{array}$$

Such a morphism $\mathcal{G}(A) \to \mathcal{F}(QA)$ does exist because the map $\mathcal{G}(QA) \to \mathcal{G}(A)$ is an isomorphism. This follows from the fact that $QA \to A$ is the adjoint of the weak equivalence $QrA \to rA$ in $\mathcal{M}$, thus a weak equivalence in $\mathcal{D}$. These morphisms form the appropriate unique natural transformation $\mathcal{G} \to \mathcal{F}(Q-)$. □

5.10. REMARK. The proof of Proposition 5.9 implies the following fact. Let $l : \mathcal{M} \rightleftarrows \mathcal{D} : r$ be a left model approximation. Even though in general the category $\mathcal{D}$ does not admit a model category structure, there is a good candidate for a "cofibrant replacement". Let us choose a cofibrant replacement Q in $\mathcal{M}$. For an object A in $\mathcal{D}$, define $QA := lQrA$ and $lQrA \to A$ to be the adjoint of the cofibrant replacement $QrA \overset{\sim}{\twoheadrightarrow} rA$. As in the case of model categories, this cofibrant replacement can be then used to construct left derived functors.

5.11. EXAMPLE. Let $l : \mathcal{M} \rightleftarrows \mathcal{D} : r$ be a left model approximation. By definition $l : \mathcal{M} \to \mathcal{D}$ is homotopy meaningful on cofibrant objects and by Proposition 5.5 the localization $\mathcal{D} \to Ho(\mathcal{D})$ exists. Thus Proposition 5.9 asserts that the total left derived functor of $l : \mathcal{M} \to \mathcal{D}$ exists. The induced functor on the homotopy categories is denoted by the same symbol $l : Ho(\mathcal{M}) \to Ho(\mathcal{D})$.

Finally let us mention that the dualization of all the material described in this section presents no difficulties and the details are left to the zealous reader. In particular there is a notion of a right model approximation. This will be briefly discussed in Section 31.

6. Spaces and small categories

In this section we recall the definitions and some basic functorial properties of two constructions which intertwine categories with spaces. To a space one can associate its so-called simplex category (see [**38**] and [**45**]). To a small category one can associate a space called its nerve.

6.1. DEFINITION. Let K be a simplicial set . The *simplex category* of K is a category, denoted by $\mathbf{K}$, whose objects are maps of the form $\sigma : \Delta[n] \to K$ and a morphism from $\sigma : \Delta[n] \to K$ to $\tau : \Delta[m] \to K$ is a commutative triangle:

$$\begin{array}{ccc} \Delta[n] & \longrightarrow & \Delta[m] \\ & {}_{\sigma}\searrow \quad \swarrow_{\tau} & \\ & K & \end{array}$$

The terminology we use comes from the fact that the objects of $\mathbf{K}$ can be identified with the simplices of K. For each n-dimensional simplex $\sigma \in K_n$, there are $n+1$ *different* morphisms called degeneracies $\{s_i : s_i\sigma \to \sigma\}_{0 \le i \le n}$ and, if $n > 0$, there are $n+1$ *different* morphisms called faces $\{d_i : d_i\sigma \to \sigma\}_{0 \le i \le n}$. Subject to the usual cosimplicial relations they generate all the morphisms in $\mathbf{K}$.

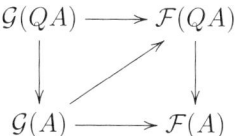

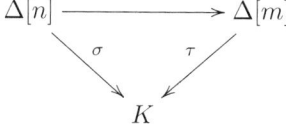

Taking the simplex category of K is natural in K and hence this process defines a functor $Spaces \to Cat$, $K \mapsto \mathbf{K}$.

6.2. EXAMPLE. The simplex category of $\Delta[0]$ is the category Δ . Thus diagrams indexed by $\mathbf{\Delta}[0]$ are cosimplicial objects. Simplex categories are therefore big and complicated. In order to simplify the situation we will put restrictions on diagrams indexed by them (see Sections 10 and 17).

Although the category $\mathbf{\Delta}[n]$ is rather complicated, it is easy to calculate colimits of diagrams indexed by it. The only non-degenerate n-dimensional simplex $\iota \in \Delta[n]$ (the simplex corresponding to the identity map $id : \Delta[n] \to \Delta[n]$) is a terminal object in the category $\mathbf{\Delta}[n]$. Thus, if $F : \mathbf{\Delta}[n] \to \mathcal{C}$ is a functor, the morphism $F(\iota) \to colim_{\mathbf{\Delta}[n]} F$ is an isomorphism.

6.3. DEFINITION. Let I be a small category. The *nerve* of I is a simplicial set $N(I)$ whose set of n-dimensional simplices is given by:

$$N(I)_n := \{i_n \stackrel{\alpha_n}{\to} \cdots \stackrel{\alpha_1}{\to} i_0 \mid \alpha_k \text{ is a morphism in } \mathcal{C}\}$$

The simplicial operators d_k and s_k are defined as follows:

$$s_0 : N(I)_0 \to N(I)_1 \ , \ s_0(i) := i \stackrel{id}{\to} i$$

Let $\sigma = (i_n \stackrel{\alpha_n}{\to} i_{n-1} \stackrel{\alpha_{n-1}}{\to} \cdots \stackrel{\alpha_1}{\to} i_0)$ be an element of $N(I)_n$. For $n > 0$ and $0 \le k \le n$, the maps $N(I)_{n+1} \stackrel{s_k}{\leftarrow} N(I)_n \stackrel{d_k}{\to} N(I)_{n-1}$ are defined as:

$$s_k(\sigma) := (i_n \stackrel{\alpha_n}{\to} \cdots \stackrel{\alpha_{k+1}}{\to} i_k \stackrel{id}{\to} i_k \stackrel{\alpha_k}{\to} \cdots \stackrel{\alpha_1}{\to} i_0)$$

$$d_k(\sigma) := \begin{cases} (i_n \stackrel{\alpha_n}{\to} \cdots \stackrel{\alpha_2}{\to} i_1) & \text{if } k = 0 \\ (i_n \stackrel{\alpha_n}{\to} \cdots \stackrel{\alpha_{k+2}}{\to} i_{k+1} \stackrel{\alpha_k \circ \alpha_{k+1}}{\longrightarrow} i_{k-1} \stackrel{\alpha_{k-1}}{\to} \cdots \stackrel{\alpha_1}{\to} i_0) & \text{if } 0 < k < n \\ (i_{n-1} \stackrel{\alpha_{n-1}}{\to} \cdots \stackrel{\alpha_1}{\to} i_0) & \text{if } k = n \end{cases}$$

Taking the nerve $N(I)$ of a category I is natural in I, i.e., this process defines a functor $N : Cat \to Spaces$.

6.4. REMARK. For any $n \ge 0$, let us denote by $[n]$ be the category given by the graph $(n \to n-1 \to \cdots \to 0)$. These categories can be assembled together to form a cosimplicial object $[-] : \Delta \to Cat$. The nerve of I can be identified with a simplicial set whose set of n-dimensional simplices is $Fun([n], I)$ and the simplicial operators are induced by the cosimplicial operators in $[-]$.

6.5. EXAMPLE. The nerve of the category $[n]$ (see Remark 6.4) coincides with the standard n-simplex $\Delta[n]$, i.e., $N([n]) = \Delta[n]$.

There are two forgetful functors $\epsilon : \mathbf{N}(I) \to I$ and $\epsilon : \mathbf{N}(I)^{op} \to I$ associated with the nerve construction:

6.6. DEFINITION. Let I be a small category.

- $\epsilon : \mathbf{N}(I) \to I$ is defined as:

$$(i_n \stackrel{\alpha_n}{\to} \cdots \stackrel{\alpha_1}{\to} i_0) = \sigma \mapsto i_0 \ , \ (s_k\sigma \stackrel{s_k}{\to} \sigma) \mapsto id_{i_0} \ , \ (d_k\sigma \stackrel{d_k}{\to} \sigma) \mapsto \begin{cases} id_{i_0} & \text{if } k > 0 \\ \alpha_1 & \text{if } k = 0 \end{cases}$$

- $\epsilon : \mathbf{N}(I)^{op} \to I$ is defined as:

$$(i_n \overset{\alpha_n}{\to} \cdots \overset{\alpha_1}{\to} i_0) = \sigma \mapsto i_n\ ,\ (s_k\sigma \overset{s_k}{\to} \sigma) \mapsto id_{i_n}\ ,\ (d_k\sigma \overset{d_k}{\to} \sigma) \mapsto \begin{cases} id_{i_n} & \text{if } k < n \\ \alpha_n & \text{if } k = n \end{cases}$$

6.7. Example. Let $f : I \to J$ be a functor of small categories. Consider the functor $N(f\downarrow -) : J \to Spaces$, which assigns to an object j the nerve of the over category $f\downarrow j$ (see Section 35) . We get a map of spaces $N(f\downarrow j) \to N(I)$ by sending $(i_n \to \cdots \to i_0, f(i_0) \to j) \in N(f\downarrow j)$ to $(i_n \to \cdots \to i_0) \in N(I)$. This map fits into the following pull-back square in Cat:

$$\begin{array}{ccccc} \mathbf{N}(f\downarrow j) & \xrightarrow{\epsilon} & f\downarrow j & \longrightarrow & J\downarrow j \\ \downarrow & & & & \downarrow \\ \mathbf{N}(I) & \xrightarrow{\epsilon} & I & \xrightarrow{f} & J \end{array}$$

where $\epsilon : \mathbf{N}(I) \to I$ and $\epsilon : \mathbf{N}(f\downarrow i) \to f\downarrow i$ are the forgetful functors. In particular this implies that $\mathbf{N}(f\downarrow j)$ can be identified with the over category $(f \circ \epsilon)\downarrow j$.

The maps $N(f\downarrow j) \to N(I)$ form a natural transformation from the diagram $N(f\downarrow -) : J \to Spaces$ to the space $N(I)$. This natural transformation satisfies the universal property of the colimit of $N(f\downarrow -)$ and hence the induced map $colim_J N(f\downarrow -) \to N(I)$ is an isomorphism.

In the case f is the identity functor $id : I \to I$, the simplex category $\mathbf{N}(I\downarrow i)$ can be identified with the over category $\epsilon\downarrow i$, where $\epsilon : \mathbf{N}(I) \to I$ is the forgetful functor. Moreover the natural maps $N(I\downarrow i) \to N(I)$ satisfy the universal property of the colimit of the diagram $N(I\downarrow -) : I \to Spaces$, and hence they induce an isomorphism $colim_I N(I\downarrow -) \to N(I)$.

Here is a list of some basic functorial properties of the nerve functor and the simplex category functor.

6.8. The nerve $N : Cat \to Spaces$ has a left adjoint $c : Spaces \to Cat$ (see [**38**]):

$$mor_{Spaces}\big(K, N(I)\big) = Fun\big(c(K), I\big)$$

Hence the nerve converts pull-backs in Cat into pull-backs in $Spaces$. In particular $N(I \times J) = N(I) \times N(J)$.

6.9. The simplex category functor $Spaces \to Cat$, $K \mapsto \mathbf{K}$, has a right adjoint $S : Cat \to Spaces$ (see [**38**]):

$$Fun(\mathbf{K}, I) = mor_{Spaces}\big(K, S(I)\big)$$

This right adjoint can be defined as $S(I) := Fun(\mathbf{\Delta}[-], I)$. It follows that the functor $Spaces \to Cat$ converts colimits in $Spaces$ into colimits in Cat.

6.10. By a direct verification one can show that the functor $Spaces \to Cat$, $K \mapsto \mathbf{K}$, converts pull-backs in $Spaces$ into pull-backs in Cat. Thus in particular the simplex category of the product of two spaces $K \times N$ can be identified with the pull-back $\mathbf{K} \times_{\mathbf{\Delta}[0]} \mathbf{N} = lim(\mathbf{K} \to \mathbf{\Delta}[0] \leftarrow \mathbf{N})$ in Cat. Nevertheless the functor $Spaces \to Cat$ does not have a left adjoint. If it had one, it would convert products in $Spaces$ into products in Cat. However, since the simplex category of $\Delta[0]$ is not a terminal object in Cat, the product of simplex categories is much bigger than the simplex category of the product. We use therefore the following notation:

Notation. The product of the simplex categories of K and N is denoted by $\mathbf{K}\tilde{\times}\mathbf{N}$, whereas it follows from our convention that the simplex category of the product $K \times N$ is denoted by $\mathbf{K} \times \mathbf{N}$.

6.11. The composite $Spaces \to Cat \xrightarrow{N} Spaces$, $K \mapsto N(\mathbf{K})$, has a right adjoint $Ex : Spaces \to Spaces$

$$mor_{Spaces}\big(N(\mathbf{K}), L\big) = mor_{Spaces}\big(K, Ex(L)\big)$$

The functor Ex can be defined as $Ex(L) := mor_{Spaces}\big(N(\mathbf{\Delta}[-]), L\big)$. It follows that the functor $Spaces \to Cat \xrightarrow{N} Spaces$ commutes with colimits. Explicitly, for any diagram $H : I \to Spaces$, the natural morphisms $colim_I N(\mathbf{H}) \to N(colim_I \mathbf{H})$ and $colim_I N(\mathbf{H}^{op}) \to N\big((colim_I \mathbf{H})^{op}\big)$ are isomorphisms. Following our convention, the symbol $\mathbf{H}$ denotes the composite $I \xrightarrow{H} Spaces \to Cat$. Notice also that by 6.9, $colim_I \mathbf{H}$ is the simplex category of $colim_I H$.

7. The pull-back process and local properties

Let $f : L \to K$ be a map of spaces. We can think about f as a functor between simplex categories and hence consider the pull-back process along f (see Section 36). By definition it is a functor $f^* : Fun(\mathbf{K}, \mathcal{C}) \to Fun(\mathbf{L}, \mathcal{C})$ which assigns to a diagram $F : \mathbf{K} \to \mathcal{C}$ the composite $\mathbf{L} \xrightarrow{f} \mathbf{K} \xrightarrow{F} \mathcal{C}$. The pull-back process commutes with compositions of maps, i.e., for any $N \xrightarrow{h} L \xrightarrow{f} K$, $(f \circ h)^*$ coincides with $h^* \circ f^*$.

If a map $f : L \to K$ is fixed, we often denote the pull-back of a diagram $F : \mathbf{K} \to \mathcal{C}$ along f by the same symbol $F : \mathbf{L} \to \mathcal{C}$.

In this paper we are particularly interested in those properties of diagrams indexed by simplex categories which are preserved by the process of pulling-back along maps of spaces.

7.1. Definition. We say that a property of diagrams indexed by simplex categories is *local* if the following statements are equivalent:

- A diagram $F : \mathbf{K} \to \mathcal{C}$ has this property.
- For any simplex $\Delta[n] \to K$, the composite $\mathbf{\Delta}[n] \to \mathbf{K} \xrightarrow{F} \mathcal{C}$ has this property.

Local properties are preserved by the pull-back process. If $F : \mathbf{K} \to \mathcal{C}$ satisfies some local property, then, for any map $f : L \to K$, so does $f^*F : \mathbf{L} \to \mathcal{C}$. Local properties are *faithfully* preserved by epimorphisms. This means that if $f : L \to K$ is an epimorphism then $F : \mathbf{K} \to \mathcal{C}$ satisfies some local property if and only if $f^*F : \mathbf{L} \to \mathcal{C}$ does so.

8. Colimits of diagrams indexed by spaces

Simplex categories are associated with geometric objects. We would like to take advantage of the intuition coming from this geometry to understand diagrams indexed by such categories and constructions on them.

Let $H : I \to Spaces$ be a diagram of spaces and $\{H(i) \to colim_I H\}_{i \in I}$ be a natural transformation which satisfies the universal property of the colimit of H. We want to describe functors indexed by the simplex category of this colimit. Recall that the simplex category of $colim_I H$ can be identified with $colim_I \mathbf{H}$ (see 6.9). To describe a functor $F : colim_I \mathbf{H} \to \mathcal{C}$ it is necessary and sufficient to have the following data (compare with Section 40):

1. for every object $i \in I$, a functor $F_i : \mathbf{H}(i) \to \mathcal{C}$;
2. for every morphism $\alpha : j \to i$ in I, $F_j : \mathbf{H}(j) \to \mathcal{C}$ should coincide with the composite $\mathbf{H}(j) \xrightarrow{\mathbf{H}(\alpha)} \mathbf{H}(i) \xrightarrow{F_i} \mathcal{C}$, i.e., $F_j = H(\alpha)^* F_i$.

If $F : colim_I \mathbf{H} \to \mathcal{C}$ is a diagram, then $F_i : \mathbf{H}(i) \to \mathcal{C}$ is given by the composite $\mathbf{H}(i) \to colim_I \mathbf{H} \xrightarrow{F} \mathcal{C}$.

In the case H is a push-out diagram we get:

8.1. PROPOSITION. *Let the following be a push-out square of spaces:*

$$\begin{array}{ccc} A & \longrightarrow & L \\ \downarrow & & \downarrow \\ B & \longrightarrow & N \end{array}$$

A diagram $F : \mathbf{L} \to \mathcal{C}$ is isomorphic to one of the form $\mathbf{L} \to \mathbf{N} \to \mathcal{C}$ if and only if $\mathbf{A} \to \mathbf{L} \xrightarrow{F} \mathcal{C}$ is isomorphic to one of the form $\mathbf{A} \to \mathbf{B} \to \mathcal{C}$. □

Analogously, to describe a functor $G : (colim_I \mathbf{H})^{op} \to \mathcal{C}$ it is necessary and sufficient to have the following data:

1. for every object $i \in I$, a functor $G_i : \mathbf{H}(i)^{op} \to \mathcal{C}$;
2. for every morphism $\alpha : j \to i$ in I, $G_j : \mathbf{H}(j)^{op} \to \mathcal{C}$ should coincide with the composite $\mathbf{H}(j)^{op} \xrightarrow{\mathbf{H}(\alpha)} \mathbf{H}(i)^{op} \xrightarrow{G_i} \mathcal{C}$.

If $G : (colim_I \mathbf{H})^{op} \to \mathcal{C}$ is a diagram, then $G_i : \mathbf{H}(i)^{op} \to \mathcal{C}$ is given by the composite $\mathbf{H}(i)^{op} \to (colim_I \mathbf{H})^{op} \xrightarrow{G} \mathcal{C}$.

The geometry of an indexing space can be used to calculate colimits. The following proposition can be exploited to construct colimits by induction on the cell decomposition of the indexing space. This allows then to reduce the study of colimits to understanding the effect of a cell attachment to the indexing space.

8.2. PROPOSITION. *Let $H : I \to Spaces$ be a functor.*

1. *For any $F : colim_I \mathbf{H} \to \mathcal{C}$, $colim_{colim_I \mathbf{H}} F = colim_I colim_{\mathbf{H}(i)} F_i$.*
2. *For any $G : (colim_I \mathbf{H})^{op} \to \mathcal{C}$, $colim_{(colim_I \mathbf{H})^{op}} G = colim_I colim_{\mathbf{H}(i)^{op}} G_i$*

Since the proofs are analogous we show only 1.

PROOF OF 1. Let us denote the map $H(i) \to colim_I H$ by ξ_i. It induces a map on colimits $colim_{\mathbf{H}(i)} F_i \to colim_{colim_I \mathbf{H}} F$. We show that $colim_{colim_I \mathbf{H}} F$, together with the natural transformation induced by these morphisms, satisfies the universal property of the colimit of the diagram $i \mapsto colim_{\mathbf{H}(i)} F_i$. Let us choose a compatible family of morphisms $\{h_i : colim_{\mathbf{H}(i)} F_i \to X\}_{i \in I}$. For every simplex σ in $colim_I H$, there exists $j \in I$ and $\tau \in H(j)$ such that $\xi_j(\tau) = \sigma$. Define $F(\sigma) \to X$ to be the composite:

$$F(\sigma) = F\big(\xi_j(\tau)\big) \to colim_{\mathbf{H}(j)} F_j \xrightarrow{h_j} X$$

It is easy to check that, for all $\sigma \in colim_I H$, these morphisms are well defined (they depend only on $\sigma \in colim_I H$) and they are compatible over $colim_I \mathbf{H}$. Hence they induce a morphism $colim_{colim_I \mathbf{H}} F \to X$. It is clear from the above description that this morphism is unique. □

8.3. REMARK. Let $H : I \to Spaces$ be a functor. We can perform two operations on H. We can take its colimit $colim_I H$ and the associated simplex category $colim_I \mathbf{H}$, or we can think about the values of H as simplex categories and take its Grothendieck construction $Gr_I \mathbf{H}$ (see Section 38). There is a functor connecting these two categories $Gr_I \mathbf{H} \to colim_I \mathbf{H}$. It sends an object (i, τ) to the image of τ under the map $H(i) \to colim_I H$. Propositions 8.2 and 40.2 say that this functor is cofinal with respect to taking colimits.

The following particular cases of Proposition 8.2 are of special interest:

8.4. COROLLARY.

1. *Let* $N = colim(B \leftarrow A \to L)$ *and* $F : \mathbf{N} \to \mathcal{C}$ *be a functor. Then the following is a push-out square:*

$$\begin{array}{ccc} colim_{\mathbf{A}} F & \longrightarrow & colim_{\mathbf{L}} F \\ \downarrow & & \downarrow \\ colim_{\mathbf{B}} F & \longrightarrow & colim_{\mathbf{N}} F \end{array}$$

2. *Let* $K = colim(K_0 \to K_1 \to K_2 \to \cdots)$ *where the telescope is possibly transfinite (in which case we assume that the values at limit ordinals are the canonical ones). Let* $F : \mathbf{K} \to \mathcal{C}$ *be a functor. Then:*

$$colim_{\mathbf{K}} F = colim(colim_{\mathbf{K}_0} F \to colim_{\mathbf{K}_1} F \to colim_{\mathbf{K}_2} F \to \cdots) \qquad \square$$

Calculating colimits of diagrams indexed by arbitrary small categories can always be reduced to calculating colimits of diagrams indexed by simplex categories. The following proposition can be shown by checking that, for all $i \in I$, the categories $\epsilon \downarrow i$ are non-empty and connected (see Proposition 37.1 and [**39**, Theorem IX.3.1]).

8.5. PROPOSITION. *The forgetful functors* $\epsilon : \mathbf{N}(I) \to I$ *and* $\epsilon : \mathbf{N}(I)^{op} \to I$ *(see Definition 6.6) are cofinal with respect to taking colimits; for any* $F : I \to \mathcal{C}$*, the induced morphisms* $colim_{\mathbf{N}(I)} \epsilon^* F \to colim_I F$ *and* $colim_{\mathbf{N}(I)^{op}} \epsilon^* F \to colim_I F$ *are isomorphisms.* $\square$

9. Left Kan extensions

Let $\mathcal{C}$ be a category closed under colimits. Consider a map of spaces $f : L \to K$. In addition to the pull-back process $f^* : Fun(\mathbf{K}, \mathcal{C}) \to Fun(\mathbf{L}, \mathcal{C})$, one can associate with f a functor going the "other direction" $f^k : Fun(\mathbf{L}, \mathcal{C}) \to Fun(\mathbf{K}, \mathcal{C})$. This functor is left adjoint to f^*, and called the left Kan extension along f (see Section 36 and [**39**, Section X.3]). In order to give an explicit construction of f^k we first have to decompose f into a "fiber diagram" $df : \mathbf{K} \to Spaces$. For a simplex $\sigma : \Delta[n] \to K$ define $df(\sigma)$ to be the space that fits into the following pull-back square:

$$\begin{array}{ccc} df(\sigma) & \longrightarrow & L \\ \downarrow & & \downarrow f \\ \Delta[n] & \xrightarrow{\sigma} & K \end{array}$$

The maps $df(\sigma) \to L$ form a natural transformation from the functor df to the space L. This natural transformation satisfies the universal property of the colimit of df, and hence the induced map $colim_{\mathbf{K}} df \to L$ is an isomorphism. Moreover f

can be expressed as $colim_{\sigma \in K}\big(df(\sigma) \to \Delta[dim(\sigma)]\big)$. This suggests that one should think about the diagram df as a decomposition of f into pieces lying over small parts of K.

On the level of simplex categories, for any $\sigma : \Delta[n] \to K$, $\mathbf{\Delta}[n]$ is isomorphic to $\mathbf{K}\!\downarrow\!\sigma$, $\mathbf{df}(\sigma)$ is isomorphic to $f\!\downarrow\!\sigma$, and the above diagram corresponds to the following pull-back square in Cat (see Section 35):

$$\begin{array}{ccc} f\!\downarrow\!\sigma & \longrightarrow & \mathbf{L} \\ \downarrow & & \downarrow{\scriptstyle f} \\ \mathbf{K}\!\downarrow\!\sigma & \longrightarrow & \mathbf{K} \end{array}$$

9.1. Definition. Let $f : L \to K$ be a map of spaces. The *left Kan extension* along f is a functor $f^k : Fun(\mathbf{L}, \mathcal{C}) \to Fun(\mathbf{K}, \mathcal{C})$, which assigns to $F : \mathbf{L} \to \mathcal{C}$ the diagram $f^kF : \mathbf{K} \to \mathcal{C}$ defined as $\mathbf{K} \ni \sigma \mapsto colim_{\mathbf{df}(\sigma)}F$.

The left Kan extension process does not modify colimits of diagrams (see also Proposition 36.2 (2)).

9.2. Proposition. *For any map $f : L \to K$, the following triangle commutes:*

$$\begin{array}{ccc} Fun(\mathbf{L}, \mathcal{C}) & \xrightarrow{\ f^k\ } & Fun(\mathbf{K}, \mathcal{C}) \\ & {\scriptstyle colim_{\mathbf{L}}}\searrow \quad \swarrow{\scriptstyle colim_{\mathbf{K}}} & \\ & \mathcal{C} & \end{array}$$

Proof. Let $F : \mathbf{L} \to \mathcal{C}$ be a diagram. As $L = colim_{\mathbf{K}}df$, according to Proposition 8.2 (1), $colim_{\mathbf{L}}F = colim_{colim_{\mathbf{K}}\mathbf{df}}F = colim_{\mathbf{K}}colim_{\mathbf{df}}F = colim_{\mathbf{K}}f^kF$. □

The left Kan extension and the pull-back process are closely related:

9.3. Proposition. *Let $f : L \to K$ be a map of spaces. Then the pull-back process $f^* : Fun(\mathbf{K}, \mathcal{C}) \to Fun(\mathbf{L}, \mathcal{C})$ is right adjoint to the left Kan extension functor $f^k : Fun(\mathbf{L}, \mathcal{C}) \to Fun(\mathbf{K}, \mathcal{C})$.* □

9.4. Corollary. *The left Kan extension $f^k : Fun(\mathbf{L}, \mathcal{C}) \to Fun(\mathbf{K}, \mathcal{C})$ commutes with colimits. Explicitly, for any diagram $\Phi : I \to Fun(\mathbf{L}, \mathcal{C})$, the natural transformation $colim_I(f^k\Phi_i) \to f^k(colim_I\Phi)$ is an isomorphism.*

Proof. This corollary is a direct consequence of f^k being a left adjoint. We can also argue directly as follows. Let $\sigma \in K$ be a simplex. By definition we have:

$$(colim_I f^k\Phi_i)(\sigma) = colim_I(f^k\Phi_i(\sigma)) = colim_{i\in I}colim_{\mathbf{df}(\sigma)}\Phi_i.$$

We can conclude that $(colim_I f^k\Phi_i)(\sigma) = colim_{\mathbf{df}(\sigma)}(colim_I\Phi) = f^k(colim_I\Phi)(\sigma)$ as the colimit functor commutes with itself. □

9.5. Corollary. *The left Kan extension process commutes with composition of maps: for any $N \xrightarrow{h} L \xrightarrow{f} K$, we have $(f \circ h)^k = f^k \circ h^k$.*

Proof. Since f^k is left adjoint to f^* and h^k is left adjoint to h^*, the composite $f^k \circ h^k$ is left adjoint to $h^* \circ f^*$. Hence, as $h^* \circ f^* = (f \circ h)^*$, the functor $f^k \circ h^k$ coincides with $(f \circ h)^k$. □

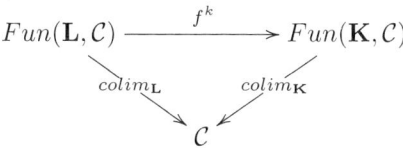

10. Bounded diagrams

In this section we introduce the notion of boundedness for diagrams indexed by simplex categories. It is a fundamental concept in this paper and plays a vital role in subsequent sections.

Since simplex categories are very complicated, in order to simplify the situation, we put restrictions on diagrams indexed by them.

10.1. DEFINITION. A functor $F : \mathbf{K} \to \mathcal{C}$ is called *bounded* if, for any degeneracy map $s_i : s_i\sigma \to \sigma$ in $\mathbf{K}$, the morphism $F(s_i) : F(s_i\sigma) \to F(\sigma)$ is an isomorphism. It is called *strongly bounded* if, for any s_i, $F(s_i) : F(s_i\sigma) \to F(\sigma)$ is the identity.

Using the simplicial identities $d_i \circ s_i = d_{i+1} \circ s_i = id$ the boundedness condition can also be expressed in terms of the boundary morphisms in $\mathbf{K}$.

10.2. PROPOSITION. *A diagram $F : \mathbf{K} \to \mathcal{C}$ is (strongly) bounded if and only if, for any simplex $\sigma \in K$ of the form $\sigma = s_i\xi$, the morphisms $F(d_i) : F(d_i\sigma) \to F(\sigma)$ and $F(d_{i+1}) : F(d_{i+1}\sigma) \to F(\sigma)$ are (identities) isomorphisms.* □

Bounded diagrams indexed by $\mathbf{K}$, with natural transformations as morphisms, form a category. This category is denoted by $Fun^b(\mathbf{K}, \mathcal{C})$. It is a full subcategory of $Fun(\mathbf{K}, \mathcal{C})$. If S denotes the set of all degeneracy morphisms in $\mathbf{K}$, then $Fun^b(\mathbf{K}, \mathcal{C})$ can be identified with $Fun(\mathbf{K}[S^{-1}], \mathcal{C})$. Diagrams indexed by such localized simplex categories were originally studied by D. W. Anderson [**1**].

The full subcategory of $Fun^b(\mathbf{K}, \mathcal{C})$ consisting of strongly bounded diagrams is denoted by $Fun^{sb}(\mathbf{K}, \mathcal{C})$. The inclusion $Fun^{sb}(\mathbf{K}, \mathcal{C}) \subset Fun^b(\mathbf{K}, \mathcal{C})$ is an equivalence as we show next.

10.3. PROPOSITION. *If $F : \mathbf{K} \to \mathcal{C}$ is a bounded diagram, then there exists an isomorphism $F \to F'$, depending functorially on F, such that F' is strongly bounded.*

PROOF. Any simplex $\sigma : \Delta[m] \to K$ can be expressed uniquely as a composite:

$$\Delta[m] \xrightarrow{s_{i_1}} \Delta[m-1] \xrightarrow{s_{i_2}} \cdots \xrightarrow{s_{i_k}} \Delta[m-k] \xrightarrow{\sigma'} K$$

where $i_1 < i_2 < \cdots < i_k$ and σ' is non-degenerate (see [**7**, Section VIII 2.3]).

Let $F : \mathbf{K} \to \mathcal{C}$ be a bounded diagram. For any $\sigma \in \mathbf{K}$, define $F'(\sigma) = F(\sigma')$ and, for any $\alpha : \sigma \to \tau$ in $\mathbf{K}$, define $F'(\alpha)$ to be the composite:

$$F(\sigma') \xrightarrow{F(\sigma \to \sigma')^{-1}} F(\sigma) \xrightarrow{F(\alpha)} F(\tau) \xrightarrow{F(\tau \to \tau')} F(\tau')$$

This assignment clearly defines a functor $F' : \mathbf{K} \to \mathcal{C}$. It is also clear that, for any degeneracy morphism s_i in $\mathbf{K}$, $F'(s_i)$ is the identity, i.e., F' is strongly bounded.

Let $F \to F'$ be the natural transformation induced by $F(\sigma) \to F(\sigma') = F'(\sigma)$. Since these morphisms are isomorphisms, the proposition has been proven. □

Boundedness is a local property (see Definition 7.1).

10.4. PROPOSITION. *A diagram $F : \mathbf{K} \to \mathcal{C}$ is bounded if and only if, for any simplex $\Delta[n] \to K$, the composite $\mathbf{\Delta}[n] \to \mathbf{K} \xrightarrow{F} \mathcal{C}$ is a bounded diagram.* □

10.5. COROLLARY. *Let $f : L \to K$ be a map of spaces. If $F : \mathbf{K} \to \mathcal{C}$ is a bounded diagram, then so is the composite $\mathbf{L} \xrightarrow{f} \mathbf{K} \xrightarrow{F} \mathcal{C}$. In this way the pull-back process along $f : L \to K$ induces a functor $f^* : Fun^b(\mathbf{K}, \mathcal{C}) \to Fun^b(\mathbf{L}, \mathcal{C})$.* □

Not only the pull-back process but also the left Kan extension preserve boundedness.

10.6. THEOREM. *Let $f : L \to K$ be a map of spaces. If $F : \mathbf{L} \to \mathcal{C}$ is a bounded diagram, then so is $f^k F : \mathbf{K} \to \mathcal{C}$. In this way the left Kan extension along f induces a functor $f^k : Fun^b(\mathbf{L}, \mathcal{C}) \to Fun^b(\mathbf{K}, \mathcal{C})$.*

A rather subtle proof of this statement is placed in Appendix A (see Theorem 33.1). It relies on a careful analysis of the degeneracy map $s_i : \Delta[n+1] \to \Delta[n]$. This is done in Section 32.

10.7. COROLLARY. *Let $f : L \to K$ be a map of spaces. The pull-back process $f^* : Fun^b(\mathbf{K}, \mathcal{C}) \to Fun^b(\mathbf{L}, \mathcal{C})$ is right adjoint to the left Kan extension functor $f^k : Fun^b(\mathbf{L}, \mathcal{C}) \to Fun^b(\mathbf{K}, \mathcal{C})$.* □

Values of a strongly bounded diagram are entirely determined by its values on the non-degenerate simplices. However, in general the category $Fun^{sb}(\mathbf{K}, \mathcal{C})$ can not be identified with $Fun(d\mathbf{K}, \mathcal{C})$, where $d\mathbf{K}$ is the full subcategory of $\mathbf{K}$ having only non-degenerate simplices as objects (see Example 10.10).

10.8. EXAMPLE. The simplex category of $\Delta[0]$ is isomorphic to Δ. Thus the category $Fun(\mathbf{\Delta}[0], \mathcal{C})$ can be identified with the category of cosimplicial objects in $\mathcal{C}$. A strongly bounded diagram indexed by $\mathbf{\Delta}[0]$ is entirely determined by the value it takes on the unique 0-dimensional simplex (the only non-degenerate one). Thus $Fun^{sb}(\mathbf{\Delta}[0], \mathcal{C})$ can be identified with $\mathcal{C}$. In this case $Fun^{sb}(\mathbf{\Delta}[0], \mathcal{C})$ coincides with the category $Fun(d\mathbf{\Delta}[0], \mathcal{C})$.

A strongly bounded diagram $F : \mathbf{\Delta}[1] \to \mathcal{C}$ is entirely determined by the pull-back diagram $\left(F(0) \overset{F(d_1)}{\longrightarrow} F(0,1) \overset{F(d_0)}{\longleftarrow} F(1)\right)$. Therefore $Fun^{sb}(\mathbf{\Delta}[1], \mathcal{C})$ is isomorphic to the category of diagrams in $\mathcal{C}$ of the shape:

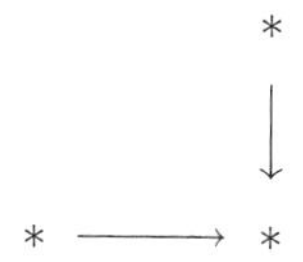

i.e., $Fun^{sb}(\mathbf{\Delta}[1], \mathcal{C})$ coincides with the category of almost square diagrams in $\mathcal{C}$ with the initial object missing. Again $Fun^{sb}(\mathbf{\Delta}[1], \mathcal{C})$ and $Fun(d\mathbf{\Delta}[1], \mathcal{C})$ can be identified.

The category $Fun^{sb}(\mathbf{\Delta}[2], \mathcal{C})$ can be identified with the category of diagrams in $\mathcal{C}$ of the shape:

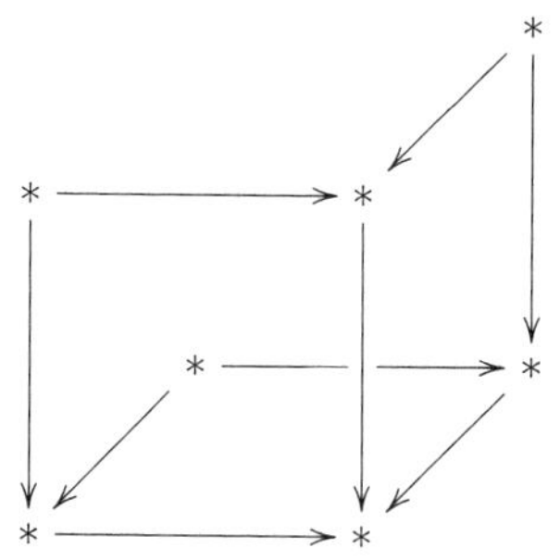

i.e., $Fun^{sb}(\mathbf{\Delta}[2], \mathcal{C})$ coincides with the category of almost 3-dimensional cubical diagrams in $\mathcal{C}$ with the initial object missing. In general, for all n, $Fun^{sb}(\mathbf{\Delta}[n], \mathcal{C})$

can be identified with the category of almost $(n+1)$-dimensional cubical diagrams in $\mathcal{C}$ with the initial object missing. For all n, $Fun^{sb}(\mathbf{\Delta}[n], \mathcal{C})$ is isomorphic to $Fun(d\mathbf{\Delta}[n], \mathcal{C})$.

10.9. EXAMPLE. Let us depict the space $\Delta[2,1]$ and its non-degenerate simplices as:

$$0 \xrightarrow{(0,1)} 1 \xrightarrow{(1,2)} 2.$$

A strongly bounded diagram $F : \mathbf{\Delta}[2,1] \to \mathcal{C}$ is determined by the following data:

$$F(0) \xrightarrow{F(d_1)} F(0,1) \xleftarrow{F(d_0)} F(1) \xrightarrow{F(d_1)} F(1,2) \xleftarrow{F(d_0)} F(2)$$

Therefore $Fun^{sb}(\mathbf{\Delta}[2,1], \mathcal{C})$ can be identified with the category of diagrams in $\mathcal{C}$ of the shape $* \to * \leftarrow * \to * \leftarrow *$. In this case $Fun^{sb}(\mathbf{\Delta}[2,1], \mathcal{C})$ coincides with $Fun(d\mathbf{\Delta}[2,1], \mathcal{C})$.

10.10. EXAMPLE. For $n > 0$, let $S^n := \Delta[n]/\partial\Delta[n]$. A strongly bounded diagram $F : \mathbf{S}^1 \to \mathcal{C}$ is determined by the data:

$$F(0) \underset{F(d_1)}{\overset{F(d_0)}{\rightrightarrows}} F(0,1)$$

Thus $Fun^{sb}(\mathbf{S}^1, \mathcal{C})$ can be identified with the category of diagrams in $\mathcal{C}$ of the shape $* \rightrightarrows *$. In this case $Fun^{sb}(\mathbf{S}^1, \mathcal{C})$ is again the same as $Fun(d\mathbf{S}^1, \mathcal{C})$.

Let 0 be the only zero-dimensional simplex in S^n, τ be the only non-degenerate n-dimensional simplex in S^n and $\alpha : 0 \to \tau$ be any morphism in $\mathbf{S}^n$. Let $F : \mathbf{S}^n \to \mathcal{C}$ be strongly bounded. If $n > 1$, using the simplicial identities, one can show that the morphism $F(\alpha) : F(0) \to F(\tau)$ does not depend on the choice of α. This morphism determines the entire diagram F. Thus, for $n > 1$, $Fun^{sb}(\mathbf{S}^n, \mathcal{C})$ can be identified with the category of morphisms in $\mathcal{C}$, i.e., with the category of diagrams in $\mathcal{C}$ of the shape $* \to *$. It follows that if $F : \mathbf{S}^n \to \mathcal{C}$ is bounded, then $colim_{\mathbf{S}^n} F$ is isomorphic to $F(\tau)$. In this case $Fun^{sb}(\mathbf{S}^n, \mathcal{C})$ and $Fun(d\mathbf{S}^n, \mathcal{C})$ do NOT coincide.

10.11. EXAMPLE. Let I be a small category and $N(I)$ be its nerve. Let us consider the forgetful functor $\epsilon : \mathbf{N}(I) \to I$, $(i_n \to \cdots \to i_0) \mapsto i_0$ (see Definition 6.6). Precomposing with this functor yields an inclusion $\epsilon^* : Fun(I, \mathcal{C}) \to Fun^b(\mathbf{N}(I), \mathcal{C})$ by which diagrams indexed by I pull-back to *bounded* diagrams indexed by the nerve $\mathbf{N}(I)$. This functor has a left adjoint $\epsilon^k : Fun^b(\mathbf{N}(I), \mathcal{C}) \to Fun(I, \mathcal{C})$, which is the restriction of the left Kan extension $\epsilon^k : Fun(\mathbf{N}(I), \mathcal{C}) \to Fun(I, \mathcal{C})$ (see Section 36).

Observe that the pull-back $\epsilon^* F$ of a functor $F : I \to \mathcal{C}$ is a bounded diagram of a special kind. Not only all the degeneracy morphisms $s_i : s_i\sigma \to \sigma$ in $N(I)$ but also the boundary morphisms $d_i : d_i\sigma \to \sigma$ for $i > 0$ are sent to isomorphisms.

10.12. EXAMPLE. Let $f : L \to K$ be an arbitrary map of spaces. The diagram $df : \mathbf{K} \to \mathit{Spaces}$ (see Section 9) is never bounded.

CHAPTER II

Homotopy theory of diagrams

11. Statements of the main results

In this section we are going to state our main results. Let I be a small category and $l : \mathcal{M} \rightleftarrows \mathcal{C} : r$ be a left model approximation. We are going to use the same symbols l and r to denote the induced functors at the level of functor categories $l : Fun(I, \mathcal{M}) \rightleftarrows Fun(I, \mathcal{C}) : r$. Recall that $\epsilon : \mathbf{N}(I) \to I$ denotes the forgetful functor (see Definition 6.6), $\epsilon^* : Fun(I, \mathcal{M}) \to Fun^b(\mathbf{N}(I), \mathcal{M})$ the pull-back process along ϵ, and $\epsilon^k : Fun^b(\mathbf{N}(I), \mathcal{M}) \to Fun(I, \mathcal{M})$ its left adjoint (the restriction of the left Kan extension of ϵ, see Example 10.11).

11.1. DEFINITION. The pair of adjoint functors:

$$Fun^b(\mathbf{N}(I), \mathcal{M}) \underset{\epsilon^* \circ r}{\overset{l \circ \epsilon^k}{\rightleftarrows}} Fun(I, \mathcal{C})$$

is called the *Bousfield-Kan approximation* of $Fun(I, \mathcal{C})$.

11.2. THEOREM. *Let K and L be simplicial sets, $f : L \to K$ be a map, and $\mathcal{M}$ be a model category.*

1. *The category $Fun^b(\mathbf{K}, \mathcal{M})$, of bounded diagrams indexed by K, can be given a model category structure where weak equivalences (respectively fibrations) are the objectwise weak equivalence (respectively fibrations).*
2. *The functor $colim_{\mathbf{K}} : Fun^b(\mathbf{K}, \mathcal{M}) \to \mathcal{M}$ is homotopy meaningful on cofibrant objects. Moreover it converts (acyclic) cofibrations in $Fun^b(\mathbf{K}, \mathcal{M})$ into (acyclic) cofibrations in $\mathcal{M}$. In particular if $F : \mathbf{K} \to \mathcal{M}$ is a cofibrant object in $Fun^b(\mathbf{K}, \mathcal{M})$, then so is $colim_{\mathbf{K}} F$ in $\mathcal{M}$.*
3. *The functor $f^k : Fun^b(\mathbf{L}, \mathcal{M}) \to Fun^b(\mathbf{K}, \mathcal{M})$ is homotopy meaningful on cofibrant objects. Moreover it converts (acyclic) cofibrations in $Fun^b(\mathbf{L}, \mathcal{M})$ into (acyclic) cofibrations in $Fun^b(\mathbf{K}, \mathcal{M})$. In particular if $F : \mathbf{L} \to \mathcal{M}$ is a cofibrant object in $Fun^b(\mathbf{L}, \mathcal{M})$, then so is $f^k F$ in $Fun^b(\mathbf{K}, \mathcal{M})$.*

11.3. THEOREM. *Let I and J be small categories, $f : I \to J$ be a functor, and $l : \mathcal{M} \rightleftarrows \mathcal{C} : r$ be a left model approximation.*

1. *The Bousfield-Kan approximation of $Fun(I, \mathcal{C})$ is a left model approximation.*
2. *Assume that $\mathcal{C}$ is closed under colimits. The Bousfield-Kan approximation of $Fun(I, \mathcal{C})$ is good for $colim_I : Fun(I, \mathcal{C}) \to \mathcal{C}$. In particular the total left derived functor of $colim_I$ (the homotopy colimit) exists.*
3. *Assume that $\mathcal{C}$ is closed under colimits. The Bousfield-Kan approximation of $Fun(I, \mathcal{C})$ is good for $f^k : Fun(I, \mathcal{C}) \to Fun(J, \mathcal{C})$. In particular the total left derived functor of f^k (the homotopy left Kan extension) exists.*

This chapter is entirely devoted to the proof of the above theorems. Let us indicate where to find the proofs of all the statements. Theorem 11.2 (1) is the main result (Theorem 13.1) of Section 13. In the same section we find Theorem 11.2 (3) as Proposition 13.3, and Theorem 11.2 (2) is a mere consequence of it, see Corollary 13.4. Section 16 is devoted to the proof of Theorem 11.3 (2) and Theorem 11.3 (3), whereas Theorem 11.3 (1) appears as Theorem 15.1.

Our approach is rather simple, maybe even naive, and the idea of the proofs is not difficult to understand. There is however one "malfunction" in the world of bounded diagrams: the image of a non-degenerate simplex by a map might very well be degenerate. This implies, as we will see in Example 12.8, that being a cofibrant diagram is not a local property as defined in Definition 7.1. To take care of this problem we have to introduce the notion of relatively bounded and relatively cofibrant diagrams. So as not to overwhelm the reader with technicalities we will postpone the definitions of these objects, as well as the proofs of the results where they play a role, to the end of the chapter.

12. Cofibrations

From this section on we start discussing the homotopy theoretical aspects of bounded diagrams. Let $\mathcal{M}$ be a model category. One of the main goals is to show that there is an appropriate model category structure on $Fun^b(\mathbf{K}, \mathcal{M})$. We start with the definition of a cofibration. The axioms will be verified in Section 13.

Let K be a space, $F : \mathbf{K} \to \mathcal{M}$ and $G : \mathbf{K} \to \mathcal{M}$ be diagrams, and $\Psi : F \to G$ be a natural transformation. For any $\sigma : \Delta[n] \to K$, let us pull-back Ψ along $\partial\Delta[n] \overset{i}{\hookrightarrow} \Delta[n] \overset{\sigma}{\to} K$, take the colimits, and define:

$$M_\Psi(\sigma) := colim\ \Big(\ colim_{\mathbf{\Delta}[n]}F \longleftarrow colim_{\partial\mathbf{\Delta}[n]}F \xrightarrow{colim_{\partial\mathbf{\Delta}[n]}\Psi} colim_{\partial\mathbf{\Delta}[n]}G\ \Big)$$

where we use the letter F to denote also the functors σ^*F and $(\sigma \circ i)^*F$. We can then form the following commutative diagram:

$$\begin{array}{ccccc}
colim_{\partial\mathbf{\Delta}[n]}F & \longrightarrow & colim_{\partial\mathbf{\Delta}[n]}G & & \\
\downarrow & & \downarrow & \searrow & \\
colim_{\mathbf{\Delta}[n]}F & \longrightarrow & M_\Psi(\sigma) & \longrightarrow & colim_{\mathbf{\Delta}[n]}G \\
\| & & & & \| \\
F(\sigma) & & \xrightarrow{\quad\Psi_\sigma\quad} & & G(\sigma)
\end{array}$$

Observe that in this way we get a functor $M_\Psi : \mathbf{K} \to \mathcal{M}$ and natural transformations $F \to M_\Psi$ and $M_\Psi \to G$ whose composite equals Ψ. In the case $dim(\sigma) = 0$, we have that $M_\Psi(\sigma) = F(\sigma)$ and the morphisms $F(\sigma) \to M_\Psi(\sigma) \to G(\sigma)$ coincide with $F(\sigma) \xrightarrow{id} F(\sigma) \xrightarrow{\Psi_\sigma} G(\sigma)$.

12.1. Definition. Let $\Psi : F \to G$ be a natural transformation in $Fun^b(\mathbf{K}, \mathcal{M})$.

- We say that $\Psi : F \to G$ is a *cofibration* if, for any non-degenerate simplex σ in K, the morphism $M_\Psi(\sigma) \to G(\sigma)$ is a cofibration in $\mathcal{M}$.
- Let $\emptyset : \mathbf{K} \to \mathcal{M}$ be the constant diagram whose value is the initial object $\emptyset$ in $\mathcal{M}$. We say that F is *cofibrant* if the natural transformation $\emptyset \to F$ is a cofibration.

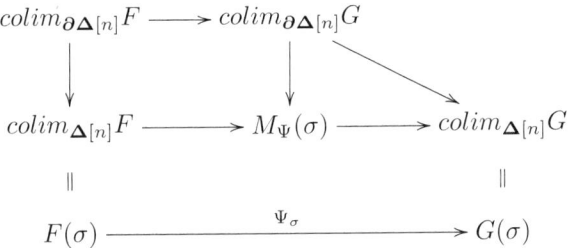

Let $\Psi : F \to G$ be a cofibration. In the case $dim(\sigma) = 0$, since the morphism $M_\Psi(\sigma) \to G(\sigma)$ coincides with $\Psi_\sigma : F(\sigma) \to G(\sigma)$, according to Definition 12.1, Ψ_σ is a cofibration. Later on we will show that in general, for any $\sigma \in K$, the morphism $\Psi_\sigma : F(\sigma) \to G(\sigma)$ is a cofibration (see Corollary 20.6 (1)). The reverse implication of course is not true.

Cofibrant diagrams can be explicitly characterized as follows:

12.2. PROPOSITION. *A bounded diagram $F : \mathbf{K} \to \mathcal{M}$ is cofibrant if and only if, for any non-degenerate simplex $\sigma : \Delta[n] \to K$, the morphism $colim_{\partial \mathbf{\Delta}[n]} F \to F(\sigma)$ is a cofibration in $\mathcal{M}$.* □

Here are some examples of cofibrant diagrams.

12.3. EXAMPLE. A diagram $F : \mathbf{\Delta}[0] \to \mathcal{M}$ in $Fun^b(\mathbf{\Delta}[0], \mathcal{M})$ is cofibrant if and only if the object $F(0)$ is cofibrant in $\mathcal{M}$.

12.4. EXAMPLE. Let X be an object in $\mathcal{M}$ which is not initial. Then the constant diagram $X : \mathbf{K} \to \mathcal{M}$ with value X is cofibrant if and only if X is cofibrant in $\mathcal{M}$ and K does not have any non-degenerate simplex of dimension 1.

12.5. EXAMPLE. A diagram $F(0) \xrightarrow{F(d_1)} F(0,1) \xleftarrow{F(d_0)} F(1)$ in $Fun^b(\mathbf{\Delta}[1], \mathcal{M})$ (cf. Example 10.8) is cofibrant if the objects $F(0)$ and $F(1)$ are cofibrant and the morphism $F(d_1) \coprod F(d_0) : F(0) \coprod F(1) \to F(0,1)$ is a cofibration.

12.6. EXAMPLE. Consider a diagram in $Fun^b(\mathbf{\Delta}[2,1], \mathcal{M})$ given by (cf. Example 10.9):

$$F(0) \xrightarrow{F(d_1)} F(0,1) \xleftarrow{F(d_0)} F(1) \xrightarrow{F(d_1)} F(1,2) \xleftarrow{F(d_0)} F(2)$$

This diagram is cofibrant if and only if, in addition to objects $F(0)$, $F(1)$, and $F(2)$ being cofibrant, the morphisms $F(d_1) \coprod F(d_0) : F(0) \coprod F(1) \to F(0,1)$ and $F(d_1) \coprod F(d_0) : F(1) \coprod F(2) \to F(1,2)$ are cofibrations. In particular a diagram $\emptyset \to B \leftarrow A \to C \leftarrow \emptyset$ is cofibrant if A is cofibrant and the morphisms $A \to B$ and $A \to C$ are cofibrations.

12.7. EXAMPLE. Let $n > 1$. A diagram $F(0) \xrightarrow{F(\alpha)} F(\tau)$ in $Fun^b(\mathbf{S}^n, \mathcal{M})$ (cf. Example 10.10) is cofibrant if $F(0)$ is cofibrant and the morphism $F(\alpha)$ is a cofibration. In particular a constant diagram in $Fun^b(\mathbf{S}^n, \mathcal{M})$, i.e., a diagram associated with the identity morphism $id : X \to X$, is cofibrant if and only if the object X is cofibrant in $\mathcal{M}$ (compare with Example 12.4).

Maps between spaces can send non-degenerate simplices to degenerate ones. Thus in general cofibrations are *not* preserved by the pull-back process. The property of a natural transformation being a cofibration is not a local property.

12.8. EXAMPLE. Consider the map $\Delta[1] \to \Delta[0]$. Let X be a cofibrant object in $\mathcal{M}$ which is not initial. The constant diagram $X : \mathbf{\Delta}[0] \to \mathcal{M}$ is clearly cofibrant in $Fun^b(\mathbf{\Delta}[0], \mathcal{M})$. However its pull-back along $\Delta[1] \to \Delta[0]$, the constant diagram $X : \mathbf{\Delta}[1] \to \mathcal{M}$, is not cofibrant in $Fun^b(\mathbf{\Delta}[1], \mathcal{M})$ (see Example 12.4).

12.9. DEFINITION. We say that a map $f : L \to K$ is *reduced* if it sends non-degenerate simplices in L to non-degenerate simplices in K.

12.10. EXAMPLE. For any map of spaces $f : L \to K$, the induced maps of nerves $N(f) : N(\mathbf{L}) \to N(\mathbf{K})$ and $N(f^{op}) : N(\mathbf{L}^{op}) \to N(\mathbf{K}^{op})$ are always reduced.

12.11. EXAMPLE. Let $f : I \to J$ be a functor. For any $j \in J$, there is a new functor $f{\downarrow}j \to I$ (see Section 35 in Appendix B). The induced map of nerves $N(f{\downarrow}j) \to N(I)$ is reduced.

12.12. PROPOSITION. *Let $\Psi : F \to G$ be a cofibration in $Fun^b(\mathbf{K}, \mathcal{M})$. If $f : L \to K$ in reduced, then the pull-back $f^*\Psi : f^*F \to f^*G$ is a cofibration in $Fun^b(\mathbf{L}, \mathcal{M})$.* □

13. $Fun^b(\mathbf{K}, \mathcal{M})$ as a model category

In this section we prove that the category of bounded diagrams, with values in a model category, forms a model category. This is Theorem 11.2 (1).

13.1. THEOREM. *Let $\mathcal{M}$ be a model category. The category $Fun^b(\mathbf{K}, \mathcal{M})$, together with the following choice of weak equivalences, fibrations, and cofibrations, satisfies the axioms of a model category:*

- *a natural transformation $\Psi : F \to G$ is a weak equivalence (respectively a fibration) if for any simplex $\sigma \in K$, $\Psi(\sigma) : F(\sigma) \to G(\sigma)$ is a weak equivalence (respectively a fibration) in $\mathcal{M}$;*
- *a natural transformation $\Psi : F \to G$ is a cofibration if it is a cofibration in the sense of Definition 12.1.*

The proof relies on the fact that the colimit behaves well with respect to cofibrations and cofibrant objects.

13.2. THEOREM. *Let $f : L \to K$ be a map and $\Psi : F \to G$ be a natural transformation in $Fun^b(\mathbf{K}, \mathcal{M})$. If Ψ is an (acyclic) cofibration, then the colimit $colim_{\mathbf{L}} f^*\Psi : colim_{\mathbf{L}} f^*F \to colim_{\mathbf{L}} f^*G$ is an (acyclic) cofibration in $\mathcal{M}$. In particular if F is cofibrant in $Fun^b(\mathbf{K}, \mathcal{M})$, then $colim_{\mathbf{L}} f^*F$ and $colim_{\mathbf{K}} F$ are cofibrant objects in $\mathcal{M}$.*

The proof of this theorem is postponed to Section 20, as it uses the techniques of relative cofibrations and reduction (see Corollary 20.5).

PROOF OF THEOREM 13.1. **MC1**, **MC2**. These axioms are obviously satisfied.

MC3. Weak equivalences and fibrations are clearly closed under retracts. Since the construction M_Ψ is natural with respect to Ψ, cofibrations are also preserved by retracts.

MC4. Let the following be a commutative square in $Fun^b(\mathbf{K}, \mathcal{M})$:

$$\begin{array}{ccc} F & \longrightarrow & E \\ {\scriptstyle\Psi}\downarrow & & \downarrow{\scriptstyle\Phi} \\ G & \longrightarrow & B \end{array}$$

where Ψ and Φ are respectively either a cofibration and acyclic fibration, or an acyclic cofibration and fibration. We need to show that in the above diagram there exists a lift $\Omega : G \to E$. We are going to construct $\Omega_\sigma : G(\sigma) \to E(\sigma)$ by induction on the dimension of σ.

Let $dim(\sigma) = 0$. Since $\Psi_\sigma : F(\sigma) \to G(\sigma)$ is an (acyclic) cofibration in $\mathcal{M}$, by the lifting axiom, there exists $h : G(\sigma) \to E(\sigma)$ which makes the following diagram commutative:

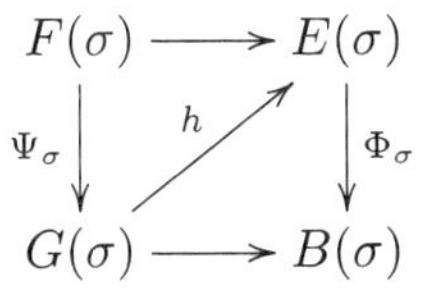

Define $\Omega_\sigma := h$.

Let us assume that the appropriate morphisms $\Omega_\sigma : G(\sigma) \to E(\sigma)$ have been constructed for all the simplices $\sigma \in K$ such that $dim(\sigma) < n$.

Let $dim(\tau) = n$. If τ is degenerate, i.e., if $\tau = s_i\xi$, we define Ω_τ as the composite:

$$G(\tau) \xrightarrow{G(s_i)} G(\xi) \xrightarrow{\Omega_\xi} E(\xi) \xrightarrow{E(s_i)^{-1}} E(\tau)$$

Using simplicial identities one can show that Ω_τ does not depend on the choice of ξ.

If τ is non-degenerate, let us consider the following commutative diagram induced by $\tau : \Delta[n] \to K$:

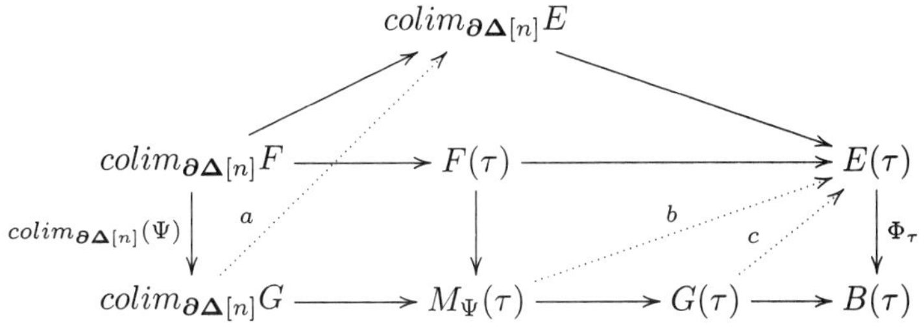

where:

- $a : colim_{\partial\mathbf{\Delta}[n]}G \to colim_{\partial\mathbf{\Delta}[n]}E$ is induced by $\{\Omega_\sigma : G(\sigma) \to E(\sigma)\}_{dim(\sigma)<n}$.
- $b : M_\Psi(\tau) \to E(\tau)$ is then constructed by the universal property of the push-out $M_\Psi(\tau)$ using the morphism a.
- Consider the case when Φ is an acyclic fibration. Since Ψ is a cofibration, by definition, $M_\Psi(\tau) \to G(\tau)$ is a cofibration in $\mathcal{M}$. We construct then the morphism $c : G(\tau) \to E(\tau)$ using b and the lifting axiom in $\mathcal{M}$.
- Consider the case when Φ is a fibration. Since Ψ is a cofibration and a weak equivalence, according to Theorem 13.2, the map $colim_{\partial\mathbf{\Delta}[n]}\Psi$ is an acyclic cofibration in $\mathcal{M}$. It follows that so is $F(\tau) \to M_\Psi(\tau)$. This implies that $M_\Psi(\tau) \to G(\tau)$ is also an acyclic cofibration. We can now construct $c : G(\tau) \to E(\tau)$ using b and the lifting axiom in $\mathcal{M}$.

We define $\Omega_\tau := c$.

The family of morphisms $\{\Omega_\sigma : G(\sigma) \to E(\sigma)\}_{\sigma\in K}$ forms a natural transformation $\Omega : G \to E$ which is the desired lift.

MC5. Let $\Psi : F \to G$ be a natural transformation in $Fun^b(\mathbf{K}, \mathcal{M})$. We need to show that Ψ can be expressed as composites $F \overset{\sim}{\hookrightarrow} F' \twoheadrightarrow G$ and $F \hookrightarrow G' \overset{\sim}{\twoheadrightarrow} G$. As in the previous case, to construct diagrams $F' : \mathbf{K} \to \mathcal{M}$, $G' : \mathbf{K} \to \mathcal{M}$ and appropriate natural transformations we argue by induction on the dimension of simplices in K.

Let $dim(\sigma) = 0$. We define $F'(\sigma)$ and $G'(\sigma)$ to be any objects that fit into the following factorizations of Ψ_σ in $\mathcal{M}$:

$$F(\sigma) \overset{\sim}{\hookrightarrow} F'(\sigma) \twoheadrightarrow G(\sigma) \quad (\text{composite } \Psi_\sigma) \qquad\qquad F(\sigma) \hookrightarrow G'(\sigma) \overset{\sim}{\twoheadrightarrow} G(\sigma) \quad (\text{composite } \Psi_\sigma)$$

Let us assume that we have constructed $F'(\sigma)$, $G'(\sigma)$, and appropriate maps for all the simplices $\sigma \in K$ whose dimension is less than n.

Let $dim(\tau) = n$. If τ is degenerate, i.e., if $\tau = s_i\xi$, we define $F'(\tau) := F'(\xi)$ and $G'(\tau) := G'(\xi)$.

Assume now that $\tau : \Delta[n] \to K$ is non-degenerate and consider the composite $\partial\Delta[n] \hookrightarrow \Delta[n] \xrightarrow{\tau} K$. By Theorem 13.2 and the inductive assumption $colim_{\partial\Delta[n]}F \overset{\sim}{\hookrightarrow} colim_{\partial\Delta[n]}F'$ and $colim_{\partial\Delta[n]}F \hookrightarrow colim_{\partial\Delta[n]}G'$ are respectively an acyclic cofibration and a cofibration.

Let:

$$M_1 := colim\big(colim_{\partial\Delta[n]}F' \overset{\sim}{\hookleftarrow} colim_{\partial\Delta[n]}F \to F(\tau)\big)$$

$$M_2 := colim\big(colim_{\partial\Delta[n]}G' \hookleftarrow colim_{\partial\Delta[n]}F \to F(\tau)\big)$$

This data can be arranged into commutative diagrams:

$$\begin{array}{ccccc}
colim_{\partial\Delta[n]}F & \longrightarrow & F(\tau) & & \\
\downarrow\wr & & \downarrow\wr & & \\
colim_{\partial\Delta[n]}F' & \longrightarrow & M_1 & & \\
\downarrow & & \downarrow & & \\
colim_{\partial\Delta[n]}G & \longrightarrow & M_\Psi(\tau) & \longrightarrow & G(\tau)
\end{array}
\qquad
\begin{array}{ccccc}
colim_{\partial\Delta[n]}F & \longrightarrow & F(\tau) & & \\
\downarrow & & \downarrow & & \\
colim_{\partial\Delta[n]}G' & \longrightarrow & M_2 & & \\
\downarrow & & \downarrow & & \\
colim_{\partial\Delta[n]}G & \longrightarrow & M_\Psi(\tau) & \longrightarrow & G(\tau)
\end{array}$$

Define $F'(\tau)$ and $G'(\tau)$ to be any objects that fit into the following factorizations of the morphisms $M_1 \to G(\tau)$ and $M_2 \to G(\tau)$:

$$M_1 \overset{\sim}{\hookrightarrow} F'(\tau) \twoheadrightarrow G(\tau) \quad , \quad M_2 \hookrightarrow G'(\tau) \overset{\sim}{\twoheadrightarrow} G(\tau)$$

In this way we get bounded diagrams $F' : \mathbf{K} \to \mathcal{M}$, $G' : \mathbf{K} \to \mathcal{M}$, and the desired natural transformations $F \overset{\sim}{\hookrightarrow} F' \twoheadrightarrow G$, $F \hookrightarrow G' \overset{\sim}{\twoheadrightarrow} G$. □

Let $f : L \to K$ be a map. We have associated with f a pair of adjoint functors: the pull-back process $f^* : Fun^b(\mathbf{K}, \mathcal{M}) \to Fun^b(\mathbf{L}, \mathcal{M})$ and the left Kan extension $f^k : Fun^b(\mathbf{L}, \mathcal{M}) \to Fun^b(\mathbf{K}, \mathcal{M})$ (see Corollary 10.7). Clearly f^* converts (acyclic) fibrations in $Fun^b(\mathbf{K}, \mathcal{M})$ into (acyclic) fibrations in $Fun^b(\mathbf{L}, \mathcal{M})$. By adjointness and K. Brown's lemma (see Proposition 3.4) this implies the following proposition, which is Theorem 11.2 (3).

13.3. PROPOSITION. *Let $f : L \to K$ be a map of spaces.*

1. *The left Kan extension $f^k : Fun^b(\mathbf{L}, \mathcal{M}) \to Fun^b(\mathbf{K}, \mathcal{M})$ converts (acyclic) cofibrations in $Fun^b(\mathbf{L}, \mathcal{M})$ into (acyclic) cofibrations in $Fun^b(\mathbf{K}, \mathcal{M})$.*
2. *The left Kan extension $f^k : Fun^b(\mathbf{L}, \mathcal{M}) \to Fun^b(\mathbf{K}, \mathcal{M})$ is homotopy meaningful on cofibrant objects.* □

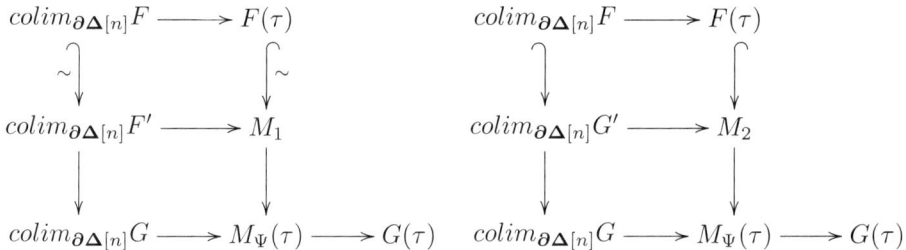

A pair of adjoint functors which satisfies the properties given in Proposition 13.3 is said to form a Quillen pair (see [**18**, Definition 4.1]). Thus we can conclude that any map $f : L \to K$ yields a Quillen pair $f^k : Fun^b(\mathbf{L}, \mathcal{M}) \rightleftarrows Fun^b(\mathbf{K}, \mathcal{M}) : f^*$.

As a particular case of Proposition 13.3, we get Theorem 11.2 (2):

13.4. COROLLARY. *The colimit functor* $colim_{\mathbf{K}} : Fun^b(\mathbf{K}, \mathcal{M}) \to \mathcal{M}$ *is homotopy meaningful on cofibrant objects. Moreover if* F *is cofibrant in* $Fun^b(\mathbf{K}, \mathcal{M})$, *then* $colim_{\mathbf{K}} F$ *is cofibrant in* $\mathcal{M}$. □

14. Ocolimit of bounded diagrams

In this section we discuss the total left derived functor of the colimit of bounded diagrams.

14.1. DEFINITION. Let $\mathcal{M}$ be a model category. The total left derived functor of $colim_{\mathbf{K}} : Fun^b(\mathbf{K}, \mathcal{M}) \to \mathcal{M}$ is denoted by $ocolim_{\mathbf{K}} : Fun^b(\mathbf{K}, \mathcal{M}) \to Ho(\mathcal{M})$.

14.2. PROPOSITION. *The functor* $ocolim_{\mathbf{K}} : Fun^b(\mathbf{K}, \mathcal{M}) \to Ho(\mathcal{M})$ *exists. It can be constructed by choosing a cofibrant replacement* Q *in* $Fun^b(\mathbf{K}, \mathcal{M})$ *and assigning to a diagram* $F \in Fun^b(\mathbf{K}, \mathcal{M})$ *the colimit* $colim_{\mathbf{K}} QF \in Ho(\mathcal{M})$. *The natural transformation* $ocolim_{\mathbf{K}} \to colim_{\mathbf{K}}$ *is induced by* $colim_{\mathbf{K}}(QF \overset{\sim}{\twoheadrightarrow} F)$.

PROOF. This is a direct consequence of Proposition 3.5 and the fact that $colim_{\mathbf{K}} : Fun^b(\mathbf{K}, \mathcal{M}) \to \mathcal{M}$ is homotopy meaningful on cofibrant objects (see Corollary 13.4). □

What is the intuition behind the construction of the ocolimit of a bounded diagram? It should be seen as a homotopy meaningful process which involves three basic steps: coproducts, push-outs, and telescopes. Let $F : \mathbf{K} \to \mathcal{M}$ be a bounded diagram. We can build its ocolimit by induction on the cell decomposition of K. For all 0-dimensional simplices σ we take cofibrant replacements $QF(\sigma) \overset{\sim}{\twoheadrightarrow} F(\sigma)$ and sum them up: $\coprod QF(\sigma)$. We then go on by attaching generalized cells along their boundaries. Let $\tau : \Delta[n] \to K$ be a non-degenerate simplex. Assume that we already know how to construct the ocolimit of F on a subcomplex $N \hookrightarrow K$ containing the boundary of τ. We then turn the morphism $colim_{\boldsymbol{\partial\Delta}[n]} QF \to F(\tau)$ into a cofibration $colim_{\boldsymbol{\partial\Delta}[n]} QF \hookrightarrow QF(\tau) \overset{\sim}{\twoheadrightarrow} F(\tau)$ and glue a generalized cell $QF(\tau)$ to $ocolim_{\mathbf{N}} F$ along its boundary $colim_{\boldsymbol{\partial\Delta}[n]} QF$; we take the push-out:

$$ocolim_{(\mathbf{N} \cup_{\boldsymbol{\partial\Delta}[n]} \boldsymbol{\Delta}[n])} F = colim\big(ocolim_{\mathbf{N}} F \leftarrow colim_{\boldsymbol{\partial\Delta}[n]} QF \hookrightarrow QF(\tau)\big).$$

This push-out process is homotopy meaningful since the objects involved are cofibrant and the morphism $colim_{\boldsymbol{\partial\Delta}[n]} QF \hookrightarrow QF(\tau)$ is a cofibration. In the case K is infinite dimensional we finish the construction by taking the telescope.

14.3. REMARK. Recall that $hocolim_{\mathbf{K}}$ denotes the total left derived functor of $colim_{\mathbf{K}} : Fun(\mathbf{K}, \mathcal{M}) \to \mathcal{M}$ (see Definition 4.1). Its construction will be given in Section 16. Let $F : \mathbf{K} \to \mathcal{M}$ be a bounded diagram. We can perform two constructions on F. Take its ocolimit $ocolim_{\mathbf{K}} F$ or take its hocolimit $hocolim_{\mathbf{K}} F$. These two constructions are both homotopy meaningful and map naturally to the colimit. However, in general $ocolim_{\mathbf{K}} F$ is NOT equivalent to $hocolim_{\mathbf{K}} F$. Consider for example the constant diagram $\Delta[0] : \mathbf{S}^2 \to Spaces$ with value $\Delta[0]$. Since it is cofibrant in $Fun^b(\mathbf{S}^2, Spaces)$, $ocolim_{\mathbf{S}^2} \Delta[0] \simeq colim_{\mathbf{S}^2} \Delta[0] = \Delta[0]$ (cf. Examples 10.10 and 12.7). On the other hand $hocolim_{\mathbf{S}^2} \Delta[0]$ is weakly equivalent to the classifying space of the category $\mathbf{S}^2$. Therefore $hocolim_{\mathbf{S}^2} \Delta[0] \simeq S^2$.

The main property that distinguishes the hocolimit from the ocolimit is the additivity with respect to the indexing spaces. By the same arguments as in Proposition 27.3 one can show that, for any functor of spaces $H : I \to Spaces$ and for any bounded diagram $F : colim_I \mathbf{H} \to \mathcal{M}$ defined over the simplex category of $colim_I H$, the morphism $colim_I hocolim_{\mathbf{H}} F \to hocolim_{colim_I \mathbf{H}} F$ is an isomorphism. Thus hocolimit is additive with respect to the indexing spaces. The ocolimit functor does not have this property (as shown by the above example since $S^2 = colim(* \leftarrow \partial\Delta[2] \hookrightarrow \Delta[2])$). This is due to the fact that absolute cofibrations are not invariant under the pull-back process (cf. Example 12.8).

14.4. REMARK. Let us assume that $\mathcal{M}$ has a functorial factorization of morphisms into cofibrations followed by acyclic fibrations. This functorial factorization can be used to construct a functorial cofibrant replacement of bounded diagrams $Q : Fun^b(\mathbf{K}, \mathcal{M}) \to Fun^b(\mathbf{K}, \mathcal{M})$. We can then construct a "rigid" ocolimit by taking $colim_{\mathbf{K}} Q(-) : Fun^b(\mathbf{K}, \mathcal{M}) \to \mathcal{M}$ (by rigid we mean a functor with values in the category $\mathcal{M}$ rather than in its homotopy category $Ho(\mathcal{M})$). The natural transformation $colim_{\mathbf{K}} Q(-) \to colim_{\mathbf{K}}$ is induced by $colim_{\mathbf{K}}(QF \overset{\sim}{\twoheadrightarrow} F)$. The functor $ocolim_{\mathbf{K}}$ coincides then with the composite $Fun^b(\mathbf{K}, \mathcal{M}) \xrightarrow{colim_{\mathbf{K}} Q(-)} \mathcal{M} \to Ho(\mathcal{M})$.

15. Bousfield-Kan approximation of $Fun(I, \mathcal{C})$

In this section we are going to show how to use model structures on categories of bounded diagrams to approximate the category of diagrams indexed by an arbitrary small category I. For this purpose we are going to use the forgetful functor $\epsilon : \mathbf{N}(I) \to I$ (see Definition 6.6) and the induced pair of adjoint functors $\epsilon^* : Fun(I, \mathcal{M}) \to Fun^b(\mathbf{N}(I), \mathcal{M})$ and $\epsilon^k : Fun^b(\mathbf{N}(I), \mathcal{M}) \to Fun(I, \mathcal{M})$ (see Example 10.11). The following is Theorem 11.3 (1).

15.1. THEOREM. *Let $l : \mathcal{M} \rightleftarrows \mathcal{C} : r$ be a left model approximation and I a small category. The Bousfield-Kan approximation (see Definition 11.1):*

$$Fun^b(\mathbf{N}(I), \mathcal{M}) \underset{\epsilon^* o r}{\overset{l o \epsilon^k}{\rightleftarrows}} Fun(I, \mathcal{C})$$

is a left model approximation.

The proof of this theorem relies on a certain "cofinality" type of statement:

15.2. LEMMA. *Let I be a small category with a terminal object denoted by t. Let $F : \mathbf{N}(I) \to \mathcal{M}$ be a bounded and cofibrant diagram. Assume that there exists $F' : I \to \mathcal{M}$ and a weak equivalence $F \overset{\sim}{\to} \epsilon^* F'$ in $Fun^b(\mathbf{N}(I), \mathcal{M})$. Then the morphism:*

$$colim_{\mathbf{N}(I)} F \to colim_{\mathbf{N}(I)} \epsilon^* F' = colim_I F' = F'(t)$$

is a weak equivalence in $\mathcal{M}$.

We postpone the proof of the lemma to Section 23 (Corollary 23.6) as it uses the techniques of relative cofibrations and reduction.

PROOF OF THEOREM 15.1. Conditions 1 and 2 of Definition 5.1 are clearly satisfied.

To show that condition 3 is satisfied we need to prove that the composite:

$$Fun^b(\mathbf{N}(I), \mathcal{M}) \xrightarrow{\epsilon^k} Fun(I, \mathcal{M}) \xrightarrow{l} Fun(I, \mathcal{C})$$

is homotopy meaningful on cofibrant objects. Let F and G be cofibrant diagrams in $Fun^b(\mathbf{N}(I), \mathcal{M})$ and $\Psi : F \xrightarrow{\sim} G$ be a weak equivalence. By definition $\epsilon^k\Psi$ assigns to $i \in I$ the following morphism in $\mathcal{M}$ (see Section 36):

$$colim_{\epsilon\downarrow i}\Psi : colim_{\epsilon\downarrow i}F \to colim_{\epsilon\downarrow i}G$$

The category $\epsilon\downarrow i$ can be identified with the simplex category $\mathbf{N}(I\downarrow i)$. Under this identification the functor $\epsilon\downarrow i \to \mathbf{N}(I)$ corresponds to the map $N(I\downarrow i) \to N(I)$ (cf. Example 6.7). Since this map is reduced (it sends non-degenerate simplices to non-degenerate ones, cf. Definition 12.9), the composites $\mathbf{N}(I\downarrow i) \to \mathbf{N}(I) \xrightarrow{F} \mathcal{M}$ and $\mathbf{N}(I\downarrow i) \to \mathbf{N}(I) \xrightarrow{G} \mathcal{M}$ are cofibrant diagrams. We can thus use Corollary 13.4 to conclude that $\epsilon^k\Psi(i) = colim_{\epsilon\downarrow i}\Psi$ is a weak equivalence between *cofibrant* objects. As $l : \mathcal{M} \to \mathcal{C}$ is homotopy meaningful on cofibrant objects, by definition, the map $l(\epsilon^k\Psi(i)) : l(\epsilon^k F(i)) \to l(\epsilon^k G(i))$ is a weak equivalence in $\mathcal{C}$.

Consider the composite:

$$Fun^b(\mathbf{N}(I), \mathcal{M}) \xleftarrow{\epsilon^*} Fun(I, \mathcal{M}) \xleftarrow{r} Fun(I, \mathcal{C})$$

To show that condition 4 of Definition 5.1 is satisfied we need to check that, for any diagram $F' : I \to \mathcal{C}$, if $F : \mathbf{N}(I) \to \mathcal{M}$ is bounded, cofibrant, and $F \to \epsilon^* rF'$ is a weak equivalence, then so is its adjoint $l\epsilon^k F \to F'$. As in the proof of condition 3, for any $i \in I$, the composite $\mathbf{N}(I\downarrow i) \to \mathbf{N}(I) \xrightarrow{F} \mathcal{M}$ is a cofibrant diagram, and therefore $\epsilon^k F(i) = colim_{N(I\downarrow i)}F$ is a cofibrant object in $\mathcal{M}$. Since the category $I\downarrow i$ has a terminal object, we can apply Lemma 15.2 to show that:

$$\epsilon^k F(i) = colim_{N(I\downarrow i)}F \to colim_{N(I\downarrow i)}\epsilon^* rF' = colim_{I\downarrow i}rF' = rF'(i)$$

is a weak equivalence. It follows that its adjoint $l\epsilon^k F(i) \to F'(i)$ is a weak equivalence in $\mathcal{C}$. □

15.3. COROLLARY. *Let $l : \mathcal{M} \rightleftarrows \mathcal{C} : r$ be a left model approximation and I a small category. The localization of $Fun(I, \mathcal{C})$ with respect to weak equivalences exists.*

PROOF. Apply Proposition 5.5. □

16. Homotopy colimits and homotopy left Kan extensions

In this section we show the second and third parts of Theorem 11.3: the Bousfield-Kan approximation is good for the colimit functor and in general for the left Kan extension.

16.1. THEOREM. *Let $\mathcal{C}$ be a category closed under colimits, $f : I \to J$ be a functor of small categories, and $l : \mathcal{M} \rightleftarrows \mathcal{C} : r$ be a left model approximation. Then the Bousfield-Kan model approximation of $Fun(I, \mathcal{C})$ is good for the functors $colim_I : Fun(I, \mathcal{C}) \to \mathcal{C}$ and $f^k : Fun(I, \mathcal{C}) \to Fun(J, \mathcal{C})$ (cf. Definition 5.8).*

PROOF. We need to show that $\epsilon^k \circ l \circ f^k : Fun^b(\mathbf{N}(I), \mathcal{M}) \to Fun(J, \mathcal{C})$ is homotopy meaningful on cofibrant objects. Since left adjoints commute with colimits they also commute with left Kan extensions and thus this functor coincides with the following composite:

$$Fun^b(\mathbf{N}(I), \mathcal{M}) \xrightarrow{\epsilon^k} Fun(I, \mathcal{M}) \xrightarrow{f^k} Fun(J, \mathcal{M}) \xrightarrow{l} Fun(J, \mathcal{C})$$

Let F and G be cofibrant diagrams in $Fun^b(\mathbf{N}(I), \mathcal{M})$ and $\Psi : F \xrightarrow{\sim} G$ be a weak equivalence. Consider the composite $\mathbf{N}(I) \xrightarrow{\epsilon} I \xrightarrow{f} J$. For any $j \in J$, the category $(f \circ \epsilon) \downarrow j$ can be identified with the simplex category $\mathbf{N}(f \downarrow j)$ (see Example 6.7). The map $N(f \downarrow j) \to N(I)$ is easily seen to be reduced (see Example 12.11). Thus according to Proposition 13.3, $\big((f \circ \epsilon)^k \Psi\big)(j) : \big((f \circ \epsilon)^k F\big)(j) \to \big((f \circ \epsilon)^k G\big)(j)$ is a weak equivalence between cofibrant objects in $\mathcal{M}$. The theorem now follows from the fact that l is homotopy meaningful on cofibrant objects. □

Even though the category $Fun(I, \mathcal{C})$ does not admit a model category structure, there is a good candidate for a "cofibrant replacement" (cf. Remark 5.10). Let us choose a cofibrant replacement Q in $Fun^b(\mathbf{N}(I), \mathcal{M})$. For any diagram $F : I \to \mathcal{C}$, define $QF := l\epsilon^k Q\epsilon^* rF$ and $QF \to F$ to be the adjoint of $Q\epsilon^* rF \overset{\sim}{\twoheadrightarrow} \epsilon^* rF$. The homotopy colimit and the homotopy left Kan extension of F can be now computed using this cofibrant replacement:

16.2. COROLLARY. *Under the same assumptions as in Theorem 16.1, the total left derived functors of $colim_I : Fun(I, \mathcal{C}) \to \mathcal{C}$ and $f^k : Fun(I, \mathcal{C}) \to Fun(J, \mathcal{C})$ exist. They can be constructed respectively by taking $colim_I QF \in Ho(\mathcal{C})$ and $f^k QF \in Ho\big(Fun(J, \mathcal{C})\big)$.* □

16.3. REMARK. When $\mathcal{M}$ has a functorial factorization of morphisms into cofibrations followed by acyclic fibrations, we can choose a functorial cofibrant replacement $Q : Fun^b(\mathbf{N}(I), \mathcal{M}) \to Fun^b(\mathbf{N}(I), \mathcal{M})$ (see Remark 14.4). This gives a functorial "cofibrant replacement" in $Fun(I, \mathcal{C})$ defined as follows: $QF := l\epsilon^k Q\epsilon^* rF$ and the map $QF \to F$ is the adjoint of $Q\epsilon^* rF \overset{\sim}{\twoheadrightarrow} \epsilon^* rF$. We can now apply this to define a "rigid" homotopy colimit and a rigid homotopy left Kan extension by taking respectively $colim_I Q(-) \in \mathcal{C}$ and $f^k QF \in Fun(J, \mathcal{C})$.

16.4. COROLLARY. *Let $\mathcal{C}$ be a category closed under colimits and $l : \mathcal{M} \rightleftarrows \mathcal{C} : r$ be its left model approximation. Then the composite:*

$$\begin{array}{ccccc} Fun(I,\mathcal{C}) & \xrightarrow{r} & Fun(I,\mathcal{M}) & \xrightarrow{\epsilon^*} & Fun^b(\mathbf{N}(I),\mathcal{M}) \\ & & & & \downarrow {\scriptstyle ocolim_{\mathbf{N}(I)}} \\ & & Ho(\mathcal{C}) & \xleftarrow{l} & Ho(\mathcal{M}) \end{array}$$

is the total left derived functor of $colim_I : Fun(I, \mathcal{C}) \to \mathcal{C}$. □

17. Relative boundedness

From this section on we introduce and discuss notions we need to prove Theorem 13.2 and Lemma 15.2. We want to warn the reader, who could be tempted to skip the end of the chapter, that it contains a very fundamental tool for the study of bounded diagrams: the reduction process (cf. Section 18).

We have seen that the pull-back process preserves boundedness (Corollary 10.5). In fact this construction preserves more properties. In order to capture this extra information we introduce in this section the notion of relative boundedness, extending Definition 10.1.

17.1. DEFINITION. Let $f : L \to K$ be a map of spaces and $F : \mathbf{L} \to \mathcal{C}$ be a functor. We say that F is *f-bounded* if, for any simplex $\sigma \in L$ such that $f(\sigma) = s_i\xi$

in K, the morphisms $F(d_i) : F(d_i\sigma) \to F(\sigma)$ and $F(d_{i+1}) : F(d_{i+1}\sigma) \to F(\sigma)$ are isomorphisms (compare with Proposition 10.2).

A simplex $\sigma \in L$ is called *f-non-degenerate* if $f(\sigma)$ is non-degenerate in K. An f-bounded diagram is determined, up to an isomorphism, by the values it takes on the f-non-degenerate simplices in L.

The full subcategory of $Fun(\mathbf{L}, \mathcal{C})$ consisting of the f-bounded diagrams is denoted by $Fun^b_f(\mathbf{L}, \mathcal{C})$. If $F : \mathbf{L} \to \mathcal{C}$ is f-bounded, then it is a bounded diagram. In this way we get an inclusion $Fun^b_f(\mathbf{L}, \mathcal{C}) \subseteq Fun^b(\mathbf{L}, \mathcal{C})$.

17.2. EXAMPLE. Let $f : L \to K$ be a map which sends non-degenerate simplices in L to non-degenerate simplices in K (such a map is called reduced, see Definition 12.9). A diagram $F : \mathbf{L} \to \mathcal{C}$ is f-bounded if and only if it is a bounded diagram, i.e., the inclusion $Fun^b_f(\mathbf{L}, \mathcal{C}) \subseteq Fun^b(\mathbf{L}, \mathcal{C})$ is an isomorphism. In particular $F : \mathbf{K} \to \mathcal{C}$ is bounded if and only if it is id_K-bounded; $Fun^b_{id}(\mathbf{K}, \mathcal{C}) = Fun^b(\mathbf{K}, \mathcal{C})$. Diagrams which are id-bounded are also called *absolutely* bounded.

17.3. EXAMPLE. Let L be a connected space and p the only map $L \to \Delta[0]$. A diagram $F : \mathbf{L} \to \mathcal{C}$ is p-bounded if and only if it is isomorphic to a constant diagram.

Relative boundedness is a local property:

17.4. PROPOSITION. *Let $f : L \to K$ be map of spaces. A diagram $F : \mathbf{L} \to \mathcal{C}$ is f-bounded if and only if, for any simplex $\sigma : \Delta[n] \to L$, the pull-back $\mathbf{\Delta}[n] \to \mathbf{L} \xrightarrow{F} \mathcal{C}$ is $(f \circ \sigma)$-bounded.* □

As a corollary we get that the relative boundedness is preserved by the pull-back process.

17.5. COROLLARY. *Let $L \xrightarrow{h} M \xrightarrow{g} K$ be maps of spaces and $F : \mathbf{M} \to \mathcal{C}$ be a diagram.*

1. *If F is g-bounded, then the pull-back $\mathbf{L} \xrightarrow{h} \mathbf{M} \xrightarrow{F} \mathcal{C}$ is $(g \circ h)$-bounded. In this way h induces a functor $h^* : Fun^b_g(\mathbf{M}, \mathcal{C}) \to Fun^b_{g \circ h}(\mathbf{L}, \mathcal{C})$.*
2. *If $h : L \to M$ is an epimorphism, then $\mathbf{L} \xrightarrow{h} \mathbf{M} \xrightarrow{F} \mathcal{C}$ is $(g \circ h)$-bounded if and only if $F : \mathbf{M} \to \mathcal{C}$ is g-bounded.* □

It follows from Corollary 17.5 that if $F : \mathbf{K} \to \mathcal{C}$ is a bounded diagram, then its pull-back $f^*F : \mathbf{L} \to \mathcal{C}$, along $f : L \to K$, is not only a bounded diagram but also f-bounded. In this way we can see that $f^* : Fun^b(\mathbf{K}, \mathcal{C}) \to Fun^b(\mathbf{L}, \mathcal{C})$ factors as $Fun^b(\mathbf{K}, \mathcal{C}) \to Fun^b_f(\mathbf{L}, \mathcal{C}) \subseteq Fun^b(\mathbf{L}, \mathcal{C})$. This extra information about f^*F is going to play an essential role.

17.6. PROPOSITION. *Let $f : L \to K$ be a map. The restriction of the left Kan extension to f-bounded diagrams $f^k : Fun^b_f(\mathbf{L}, \mathcal{C}) \to Fun^b(\mathbf{K}, \mathcal{C})$ is left adjoint to the pull-back process $f^* : Fun^b(\mathbf{K}, \mathcal{C}) \to Fun^b_f(\mathbf{L}, \mathcal{C})$.*

PROOF. First observe that since an f-bounded diagram $F : \mathbf{L} \to \mathcal{C}$ is absolutely bounded, its left Kan extension $f^kF : \mathbf{K} \to \mathcal{C}$ is also absolutely bounded (see Theorem 10.6). The proposition now follows from Corollary 10.7 and the fact that $Fun^b_f(\mathbf{L}, \mathcal{C})$ is a full subcategory of $Fun^b(\mathbf{L}, \mathcal{C})$. □

Even though absolute boundedness and to some extent relative boundedness are preserved by left Kan extensions (see Theorem 10.6 and Proposition 17.6), the relative boundedness in general does *not* have this property. Consider for example the maps $id : \Delta[0] \overset{d_1}{\hookrightarrow} \Delta[1] \overset{s_0}{\to} \Delta[0]$. Let X be an object in $\mathcal{C}$ which is not an initial one. It is clear that the constant diagram $X : \mathbf{\Delta}[0] \to \mathcal{C}$, with value X, is id-bounded. Its left Kan extension $(d_1)^k X$ however, corresponds to the diagram in $Fun^b(\mathbf{\Delta}[1], \mathcal{C})$ given by $X \overset{id}{\to} X \leftarrow \emptyset$ (cf. Example 10.8). This diagram is not s_0-bounded (see Example 17.3).

18. Reduction process

In this section we introduce a reduction process. The aim is to reduce the study of relatively bounded diagrams to the study of absolutely bounded diagrams (id-bounded).

The reduction process is motivated by the following property of the relative boundedness:

18.1. LEMMA. *Let $s_i : \Delta[n+1] \to \Delta[n]$ be the i-th degeneracy. The induced functor $s_i^* : Fun^b(\mathbf{\Delta}[n], \mathcal{C}) \to Fun^b_{s_i}(\mathbf{\Delta}[n+1], \mathcal{C})$ is an equivalence of categories. Explicitly, a diagram $F : \mathbf{\Delta}[n+1] \to \mathcal{C}$ is s_i-bounded if and only if there exists a bounded diagram $F' : \mathbf{\Delta}[n] \to \mathcal{C}$ for which the composite $\mathbf{\Delta}[n+1] \overset{s_i}{\to} \mathbf{\Delta}[n] \overset{F'}{\to} \mathcal{C}$ is isomorphic to F. Such an F' is unique up to an isomorphism.*

PROOF. The uniqueness of F' follows easily from the fact that the degeneracy $s_i : \Delta[n+1] \to \Delta[n]$ is an epimorphism.

To prove its existence, we show that F' is explicitly given by the composite $\mathbf{\Delta}[n] \overset{d_i}{\hookrightarrow} \mathbf{\Delta}[n+1] \overset{F}{\to} \mathcal{C}$. We have to construct a natural transformation $s_i^* F' \to F$ which is an isomorphism. Since both diagrams are bounded it is enough to construct isomorphisms $s_i^* F'(\sigma) \to F(\sigma)$ for the non-degenerate simplices $\sigma \in \Delta[n+1]$ (these isomorphisms should be natural with respect to σ).

Let $\sigma = (l_m > \cdots > l_0)$ be a non-degenerate simplex in $(\Delta[n+1])_m$. If σ does not contain the vertex i, then $(d_i \circ s_i)(\sigma) = \sigma$. In this case we define the morphism $s_i^* F'(\sigma) = F\big((d_i \circ s_i)(\sigma)\big) \to F(\sigma)$ to be the identity.

Assume that σ contains the vertex i. Let k be such that $l_k = i$. We consider two cases. First, assume in addition that σ contains also the vertex $i+1$, i.e., $l_{k+1} = i+1$. In this case $(d_i \circ s_i)(\sigma) = (l_m > \cdots > l_{k+2} > i = i > \cdots > l_0) \in (\Delta[n+1])_m$, and hence $(d_i \circ s_i)(\sigma) = s_k d_{k+1} \sigma$. Since $s_i(\sigma) \in \Delta[n]$ is of the form $s_k \tau$, the morphisms $F(d_k) : F(d_k \sigma) \to F(\sigma)$ and $F(d_{k+1}) : F(d_{k+1}\sigma) \to F(\sigma)$ are isomorphisms (F is s_i-bounded). We define $s_i^* F'(\sigma) \to F(\sigma)$ to be the composite:

$$s_i^* F'(\sigma) = F\big((d_i \circ s_i)(\sigma)\big) = F(s_k d_{k+1} \sigma) \xrightarrow{F(s_k)} F(d_{k+1}\sigma) \xrightarrow{F(d_{k+1})} F(\sigma)$$

It is clear that this composite is an isomorphism as $F(s_k)$ is so (F is a bounded diagram).

Assume that σ does not contain the vertex $i+1$. Consider then the simplex $\tau = (i_m > \cdots > i_{k+1} > i+1 > i_k > \cdots > l_0) \in (\Delta[n+1])_{m+1}$. The above discussion shows that $F(d_k) : F(d_k \tau) \to F(\tau)$ and $F(d_{k+1}) : F(d_{k+1}\tau) \to F(\tau)$ are isomorphisms. Observe that $(d_i \circ s_i)(\sigma) = d_k \tau$ and $\sigma = d_{k+1}\tau$. We define

$s_i^*F'(\sigma) \to F(\sigma)$ to be the composite:

$$s_i^*F'(\sigma) = F\big((d_i \circ s_i)(\sigma)\big) = F(d_k\tau) \xrightarrow{F(d_k)} F(\tau) \xrightarrow{F(d_{k+1})^{-1}} F(d_{k+1}\tau) = F(\sigma)$$

One can check that these morphisms induce the desired natural transformation $s_i^*F' \xrightarrow{\simeq} F$. □

18.2. PROPOSITION. *Consider the following commutative diagram:*

$$\begin{array}{ccccc} \Delta[n+1] & \xrightarrow{s_i} & \Delta[n] & & \\ \downarrow{\scriptstyle\tau} & & \downarrow{\scriptstyle\xi} & \searrow & \\ L & \xrightarrow{h} & M & \xrightarrow{g} & K \end{array} \qquad (f = g \circ h : L \to K)$$

where

$$\begin{array}{ccc} \Delta[n+1] & \xrightarrow{s_i} & \Delta[n] \\ \downarrow{\scriptstyle\tau} & & \downarrow{\scriptstyle\xi} \\ L & \xrightarrow{h} & M \end{array}$$

is a push-out square.

Then:

1. *The map h induces an equivalence of categories:*
$$h^* : Fun_g^b(\mathbf{M}, \mathcal{C}) \simeq Fun_f^b(\mathbf{L}, \mathcal{C})$$
Explicitly, a diagram $F : \mathbf{L} \to \mathcal{C}$ is f-bounded if and only if there exists a g-bounded diagram $F' : \mathbf{M} \to \mathcal{C}$ for which the composite $\mathbf{L} \xrightarrow{h} \mathbf{M} \xrightarrow{F'} \mathcal{C}$ is isomorphic to F. Any such F' is unique up to an isomorphism.
2. *The following diagram commutes:*

$$\begin{array}{ccc} Fun_g^b(\mathbf{M}, \mathcal{C}) & \xrightarrow{h^*} & Fun_f^b(\mathbf{L}, \mathcal{C}) \\ {\scriptstyle colim_{\mathbf{M}}}\searrow & & \swarrow{\scriptstyle colim_{\mathbf{L}}} \\ & \mathcal{C} & \end{array}$$

PROOF OF 1. Let $F : \mathbf{L} \to \mathcal{C}$ be an f-bounded diagram. It is not difficult to see that the composite $\mathbf{\Delta}[n+1] \xrightarrow{\tau} \mathbf{L} \xrightarrow{F} \mathcal{C}$ is s_i-bounded. Thus Propositions 8.1 and 18.1 imply that there exists a diagram $F' : \mathbf{M} \to \mathcal{C}$ for which the composite $\mathbf{L} \xrightarrow{h} \mathbf{M} \to \mathcal{C}$ is isomorphic to F. Since $h : L \to M$ is an epimorphism we can conclude two things. First, $F' : \mathbf{M} \to \mathcal{C}$ is g-bounded (see Corollary 17.5). Second, F' is unique up to an isomorphism.

PROOF OF 2. Let $G : \mathbf{M} \to \mathcal{C}$ be a g-bounded diagram. Since M can be expressed as a push-out $M = colim(L \xleftarrow{\tau} \Delta[n+1] \xrightarrow{s_i} \Delta[n])$, according to Corollary 8.4:

$$\begin{aligned} colim_{\mathbf{M}}G &= colim\big(colim_{\mathbf{L}}G \leftarrow colim_{\mathbf{\Delta}[n+1]}G \to colim_{\mathbf{\Delta}[n]}G\big) \\ &= colim\big(colim_{\mathbf{L}}G \leftarrow G(h(\tau)) \xrightarrow{G(s_i)} G(\xi)\big). \end{aligned}$$

The boundedness condition on G implies that $G(s_i) : G\big(h(\tau)\big) \to G(\xi)$ is an isomorphism. Hence so is $colim_{\mathbf{L}}G \to colim_{\mathbf{M}}G$. □

The inverse for h^* in Proposition 18.2 can be identified with the left Kan extension h^k (see Section 9). The proof however requires Lemma 33.3 in Appendix A.

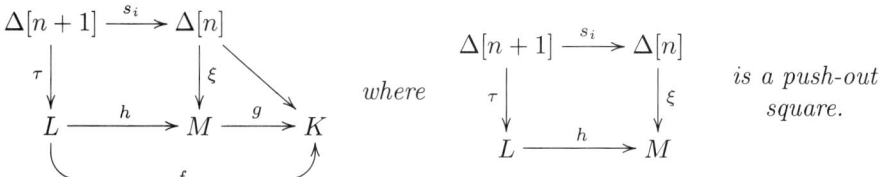

18.3. PROPOSITION. *Assume that we are in the same setting as in Proposition 18.2. Let $G : \mathbf{M} \to \mathcal{C}$ be g-bounded. Then, for any simplex $\sigma : \Delta[m] \to M$, the map $dh(\sigma) \to \Delta[m]$ induces an isomorphism $h^k h^* G(\sigma) = colim_{\mathbf{dh}(\sigma)} G \to G(\sigma)$.*

PROOF. By pulling back $\sigma : \Delta[m] \to M$ along the diagram given in Proposition 18.2 we get the following commutative cube, where all the side squares are pull-backs and the top and bottom squares are push-outs:

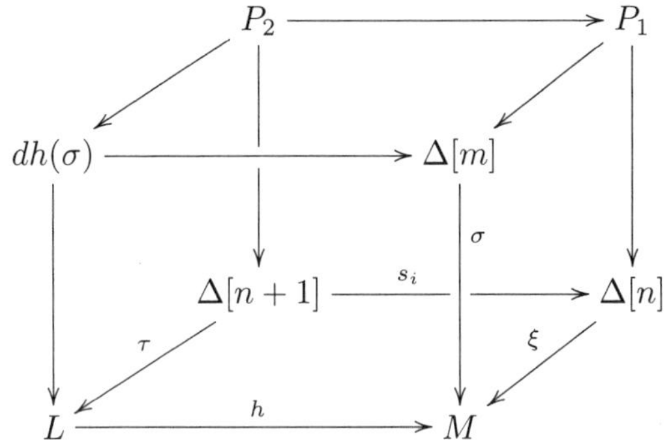

Let $G : \mathbf{M} \to \mathcal{C}$ be g-bounded. According to Corollary 8.4:

$$G(\sigma) = colim_{\Delta[m]} G = colim\big(colim_{\mathbf{dh}(\sigma)} G \leftarrow colim_{\mathbf{P_2}} G \to colim_{\mathbf{P_1}} G\big)$$

Since G is g-bounded thus in particular it is a bounded diagram. Lemma 33.3 implies therefore that $colim_{\mathbf{P_2}} G \to colim_{\mathbf{P_1}} G$ is an isomorphism. It follows that so is $colim_{\mathbf{dh}(\sigma)} G \to G(\sigma)$. □

Recall that a map of spaces is called reduced (cf. Definition 12.9) if it sends non-degenerate simplices to non-degenerate simplices. In order to study relatively bounded diagrams by looking at *absolutely* bounded diagrams, we introduce a reduction process. It is related to factoring any map in a canonical way into an epimorphism followed by a reduced map. We start with observing that reduced maps can be characterized in terms of lifting properties with respect to degeneracy maps.

18.4. PROPOSITION. *A map $f : L \to K$ is reduced if and only if in any commutative diagram of the form:*

$$\begin{array}{ccc} \Delta[n+1] & \longrightarrow & L \\ s_i \downarrow & & \downarrow f \\ \Delta[n] & \longrightarrow & K \end{array}$$

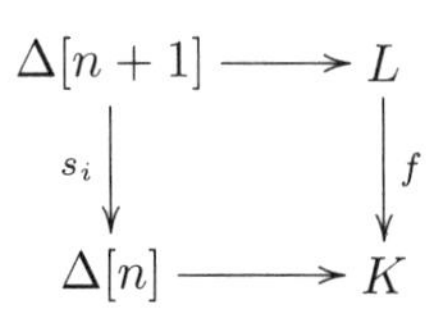

there is a lift, i.e., a map $\Delta[n] \to L$, such that the resulting diagram with five arrows commutes. Such a lift, if it exists, is necessarily unique. □

18.5. PROPOSITION. *For any map $f : L \to K$, there is a functorial factorization $red(f) = (L \xrightarrow{f_{red}} red(f) \to K)$ where:*

1. *the map $red(f) \to K$ is reduced;*
2. *if $F = (L \to X \to K)$ is another factorization of f, where $X \to K$ is reduced, then there exists a unique map of factorizations $red(f) \to F$, i.e.,*

there exists a unique map of spaces $red(f) \to X$ for which the following diagram commutes:

$$\begin{array}{ccccc} & & red(f) & & \\ & {\scriptstyle f_{red}}\nearrow & \downarrow & \searrow & \\ L & \longrightarrow & X & \longrightarrow & K \\ & & \underbrace{\hphantom{XXXXX}}_{f} & & \end{array}$$

PROOF. Let J be the set of all commutative diagrams of the form:

$$\begin{array}{ccc} \Delta[n+1] & \longrightarrow & L \\ {\scriptstyle s_i}\downarrow & & \downarrow{\scriptstyle f} \\ \Delta[n] & \longrightarrow & K \end{array}$$

If d is such a diagram, then we use the same symbol to denote the number i. Define $f_{red} : L \to red(f)$ to be the map that fits into the following push-out square:

$$\begin{array}{ccc} \coprod_{d\in J} \Delta[n+1] & \longrightarrow & L \\ {\scriptstyle \coprod_{d\in J} s_d}\downarrow & & \downarrow{\scriptstyle f_{red}} \\ \coprod_{d\in J} \Delta[n] & \longrightarrow & red(f) \end{array}$$

Observe that f_{red} is an epimorphism. Define $red(f) \to K$ to be the map induced by $f : L \to K$ and the "evaluation" map $\coprod_J \Delta[n] \to K$.

We are going to show that the factorization $L \xrightarrow{f_{red}} red(f) \to K$ satisfies conditions 1 and 2 of the proposition. Consider a commutative diagram:

$$\begin{array}{ccc} \Delta[n+1] & \longrightarrow & red(f) \\ {\scriptstyle s_i}\downarrow & & \downarrow \\ \Delta[n] & \longrightarrow & K \end{array}$$

The map $red(f) \to K$ was constructed precisely in such a way that we can find a lift in the above square in the case when $\Delta[n+1] \to red(f)$ can be factored as $\Delta[n+1] \to L \xrightarrow{f_{red}} red(f)$. Since $f_{red} : L \to red(f)$ is an epimorphism such a factorization always exists. This shows that $red(f) \to K$ is reduced.

Let $F = (L \to X \to K)$ be another factorization of f, where $X \to K$ is reduced. It follows that in any commutative diagram of the form:

$$\begin{array}{ccccc} \Delta[n+1] & \longrightarrow & L & \longrightarrow & X \\ {\scriptstyle s_i}\downarrow & & & & \downarrow \\ \Delta[n] & & \longrightarrow & & K \end{array}$$

there is always a lift. Such a lift is necessarily unique. By "summing up" over these diagrams we get a unique map of factorizations $red(f) \to F$. □

18.6. DEFINITION. The factorization $L \xrightarrow{f_{red}} red(f) \to K$ of a map $f : L \to K$, constructed in Proposition 18.5, is called the *reduction* of f.

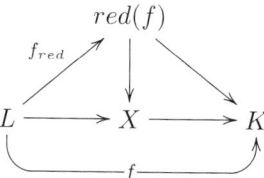

Using the reduction process, checking the condition for relative boundedness can always be reduced to checking the condition for absolute boundedness.

18.7. THEOREM. *Let $f : L \to K$ be a map and $L \xrightarrow{f_{red}} red(f) \to K$ be its reduction. Then the functors:*

$$Fun^b_f(\mathbf{L}, \mathcal{C}) \underset{(f_{red})^*}{\overset{(f_{red})^k}{\rightleftarrows}} Fun^b(\mathbf{red}(f), \mathcal{C})$$

are inverse equivalences of categories. Explicitly, $F : \mathbf{L} \to \mathcal{C}$ is an f-bounded diagram if and only if there exists a bounded diagram $F' : \mathbf{red}(f) \to \mathcal{C}$ for which the composite $\mathbf{L} \xrightarrow{f_{red}} \mathbf{red}(f) \xrightarrow{F'} \mathcal{C}$ is isomorphic to F. Any such F' is isomorphic to $(f_{red})^k F$.

PROOF. The map $h : red(f) \to K$ sends non-degenerate simplices in $red(f)$ to non-degenerate simplices in K (it is reduced). Thus a diagram is h-bounded if and only if it is a bounded diagram. The corollary follows now from Propositions 17.6 and 18.2, since to construct $f_{red} : L \to red(f)$ we glued to L maps of the form $s_i : \Delta[n+1] \to \Delta[n]$. □

19. Relative cofibrations

Let $\mathcal{M}$ be a model category. We have seen in Proposition 12.12 that the pull-back of a cofibration along a reduced map is again a cofibration. This is however no longer true as soon as the map is not reduced. In order to capture those properties of absolute cofibrations which are local, we introduce now the notion of a relative cofibration.

19.1. DEFINITION. Let $f : L \to K$ be a map and $\Psi : F \to G$ be a natural transformation in $Fun^b_f(\mathbf{L}, \mathcal{M})$. For any simplex $\sigma : \Delta[n] \to L$, pull-back Ψ along $\partial\Delta[n] \hookrightarrow \Delta[n] \xrightarrow{\sigma} L$, take colimits, and define:

$$M_\Psi(\sigma) := colim\ \Big(\ colim_{\mathbf{\Delta}[n]} F \longleftarrow colim_{\partial\mathbf{\Delta}[n]} F \xrightarrow{colim_{\partial\mathbf{\Delta}[n]} \Psi} colim_{\partial\mathbf{\Delta}[n]} G\ \Big)$$

(cf. Section 12).

- We say that $\Psi : F \to G$ is an *(acyclic) f-cofibration* if, for any simplex $\sigma \in L$ such that $f(\sigma)$ is non-degenerate in K, the morphism $M_\Psi(\sigma) \to G(\sigma)$ is an (acyclic) cofibration in $\mathcal{M}$. We also are going to use the term an (acyclic) cofibration *relative* to f to name an (acyclic) f-cofibration.
- Let $\emptyset : \mathbf{L} \to \mathcal{M}$ be the constant diagram whose value is the initial object $\emptyset$ in $\mathcal{M}$. We say that F is *f-cofibrant* if the natural transformation $\emptyset \to F$ is an f-cofibration.

19.2. REMARK. A priori it is unclear at this moment whether an acyclic cofibration, as defined in 19.1, is the same as a cofibration which is a weak equivalence. This will be shown in Corollary 20.6 (3). Until then, the term *acyclic cofibration* is always taken as in Definition 19.1, even when applied to *id*-bounded diagrams, as it will be the case in Proposition 19.7 for example.

A natural transformation $\Psi : F \to G$ is a cofibration in $Fun^b(\mathbf{K}, \mathcal{M})$ (as defined in 12.1) if it is a cofibration relative to $id : K \to K$. We will sometimes refer to such a transformation as to an *absolute* cofibration.

Diagrams that are f-cofibrant can be explicitly characterized as follows:

19.3. PROPOSITION. *Let $f : L \to K$ be a map and $F : \mathbf{L} \to \mathcal{M}$ be an f-bounded diagram. Then F is f-cofibrant if and only if, for any simplex $\sigma : \Delta[n] \to L$ such that $f(\sigma)$ is non-degenerate in K, the morphism $colim_{\partial\mathbf{\Delta}[n]}F \to F(\sigma)$ is a cofibration in $\mathcal{M}$.* □

The most significant aspect of being a relative cofibration is that this property can be checked *locally.* This is the key feature that absolute cofibrations are missing. Relative cofibrations have been introduced to enlarge the class of absolute cofibrations so the notion of cofibrancy would become local. The following is left as an easy exercise:

19.4. PROPOSITION. *Let $f : L \to K$ be a map and $\Psi : F \to G$ be a natural transformation in $Fun^b_f(\mathbf{L}, \mathcal{C})$. The following are equivalent:*

1. *Ψ is an (acyclic) f-cofibration.*
2. *For any simplex $\sigma : \Delta[n] \to L$, the pull-back of Ψ along σ is an (acyclic) $(f \circ \sigma)$-cofibration.*
3. *For any simplex $\sigma : \Delta[n] \to L$ such that $f(\sigma)$ is non-degenerate in K, the pull-back of Ψ along σ is an (acyclic) $(f \circ \sigma)$-cofibration.* □

As a corollary we get a useful procedure to check inductively on the cell decomposition of L whether a diagram $F : \mathbf{L} \to \mathcal{M}$ is f-cofibrant (the same can be used for detecting if a natural transformation is an f-cofibration). Each step of this procedure consists of:

19.5. COROLLARY. *Let $\sigma : \Delta[n] \to K$ be a simplex of K and $F : \mathbf{\Delta}[n] \to \mathcal{M}$ be a σ-bounded diagram.*

1. *Assume that σ is degenerate. Then F is σ-cofibrant if and only if, for any face map $d_i : \Delta[n-1] \to \Delta[n]$, the composite $\mathbf{\Delta}[n-1] \xrightarrow{d_i} \mathbf{\Delta}[n] \xrightarrow{F} \mathcal{M}$ is $(\sigma \circ d_i)$-cofibrant.*
2. *Assume that σ is non-degenerate. Then F is σ-cofibrant if and only if, in addition to $\mathbf{\Delta}[n-1] \xrightarrow{d_i} \mathbf{\Delta}[n] \xrightarrow{F} \mathcal{M}$ being $d_i\sigma$-cofibrant for any i, the morphism $colim_{\partial\mathbf{\Delta}[n]}F \to F(\sigma)$ is a cofibration in $\mathcal{M}$.* □

In general the way relative cofibrations behave with respect to the pull-back process can be described as follows:

19.6. COROLLARY. *Let $L \xrightarrow{h} M \xrightarrow{g} K$ be maps of spaces and $\Psi : F \to G$ be a natural transformation in $Fun^b_g(\mathbf{M}, \mathcal{M})$.*

1. *If Ψ is an (acyclic) g-cofibration, then its pull-back along h is an (acyclic) $(g \circ h)$-cofibration.*
2. *If $h : L \to M$ is an epimorphism, then Ψ is an (acyclic) g-cofibration if and only if its pull-back along h is an (acyclic) $(h \circ g)$-cofibration.* □

Corollary 19.6 implies for example that if Ψ is a cofibration in $Fun^b(\mathbf{K}, \mathcal{M})$, then its pull-back $f^*\Psi$ along $f : L \to K$ is an f-cofibration in $Fun^b_f(\mathbf{L}, \mathcal{M})$.

Using the reduction process (see Definition 18.6) checking the condition for a relative cofibration can always be reduced to checking the condition for an absolute cofibration. Theorem 18.7 and Corollary 19.6 imply:

19.7. PROPOSITION. *Let $L \xrightarrow{f_{red}} red(f) \to K$ be the reduction of $f : L \to K$. A natural transformation $\Psi : F \to G$ in $Fun^b_f(\mathbf{L}, \mathcal{M})$ is an (acyclic) f-cofibration if* *and only if* $(f_{red})^k\Psi$ *is an (acyclic) id-cofibration in* $Fun^b(\mathbf{red}(f), \mathcal{M})$. *In particular* F *is* f*-cofibrant if and only if* $(f_{red})^k F$ *is absolutely cofibrant.* $\square$

20. Cofibrations and colimits

In this section we generalize Propositions 2.5 and 2.6 to more complicated diagrams. A large part of the section is devoted to the proof of Theorem 13.2. We show in particular that the colimit converts relative (acyclic) cofibrations in $Fun^b_f(\mathbf{L}, \mathcal{M})$ into (acyclic) cofibrations in $\mathcal{M}$.

A natural transformation in $Fun^b_f(\mathbf{L}, \mathcal{M})$ is an (acyclic) f-cofibration if, for any simplex $\sigma : \Delta[n] \to L$ such that $f(\sigma)$ is non-degenerate, the colimit of its pull-back along $\partial\Delta[n] \hookrightarrow \Delta[n] \xrightarrow{\sigma} L$ satisfy a certain condition (see Definition 19.1). This information can be assembled to give a similar condition for more general maps $B \hookrightarrow A \to L$. The following proposition can be shown by induction on the dimension of the relative space (A, B), applying Proposition 2.6 (3) in the case when this dimension is infinite.

20.1. PROPOSITION. *Let* $B \hookrightarrow A \xrightarrow{g} L$ *and* $f : L \to K$ *be maps, and* $\Psi : F \to G$ *be an (acyclic)* f*-cofibration in* $Fun^b_f(\mathbf{L}, \mathcal{M})$. *Consider the following commutative square:*

$$\begin{array}{ccc} colim_{\mathbf{B}} F & \longrightarrow & colim_{\mathbf{A}} F \\ {\scriptstyle colim_{\mathbf{B}}\Psi}\downarrow & & \downarrow{\scriptstyle colim_{\mathbf{A}}\Psi} \\ colim_{\mathbf{B}} G & \longrightarrow & colim_{\mathbf{A}} G \end{array}$$

Let $M = colim\big(colim_{\mathbf{B}}G \leftarrow colim_{\mathbf{B}}F \to colim_{\mathbf{A}}F\big)$ *and* $M \to colim_{\mathbf{A}}G$ *be the morphism induced by the commutativity of the above square. If for any non-degenerate simplex* $\sigma \in A\backslash B$, $(f \circ g)(\sigma)$ *is non-degenerate in* K, *then* $M \to colim_{\mathbf{A}}G$ *is an (acyclic) cofibration in* $\mathcal{M}$. $\square$

20.2. COROLLARY. *Let* $L \hookrightarrow K$ *be a monomorphism. If* $F : \mathbf{K} \to \mathcal{M}$ *is an absolutely cofibrant diagram, then* $colim_{\mathbf{L}}F \to colim_{\mathbf{K}}F$ *is a cofibration in* $\mathcal{M}$. $\square$

We are now ready to prove the key homotopical properties of relative cofibrations:

20.3. THEOREM. *Let* $f : L \to K$ *be a map and* $\Psi : F \to G$ *be a natural transformation in* $Fun^b_f(\mathbf{L}, \mathcal{M})$.

1. *If* Ψ *is an (acyclic)* f*-cofibration, then* $colim_{\mathbf{L}}\Psi : colim_{\mathbf{L}}F \to colim_{\mathbf{L}}G$ *is an (acyclic) cofibration in* $\mathcal{M}$.
2. *If* Ψ *is an* f*-cofibration and, for any simplex* $\sigma \in L$, $\Psi_\sigma : F(\sigma) \to G(\sigma)$ *is a weak equivalence, then* $colim_{\mathbf{L}}\Psi : colim_{\mathbf{L}}F \to colim_{\mathbf{L}}G$ *is an acyclic cofibration.*

Since the proofs are analogous we show only 1. We start with discussing the absolute case.

20.4. LEMMA. *Let* $\Psi : F \to G$ *be a natural transformation in* $Fun^b(\mathbf{K}, \mathcal{M})$. *If* Ψ *is an (acyclic) id-cofibration, then* $colim_{\mathbf{K}}\Psi : colim_{\mathbf{K}}F \to colim_{\mathbf{K}}G$ *is an (acyclic) cofibration in* $\mathcal{M}$.

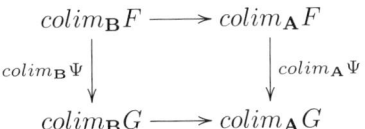

PROOF. Assume first that K is finite dimensional. In this case we prove the lemma by induction on the dimension of K.

If $dim(K) = 0$, then the morphism $colim_{\mathbf{K}}\Psi : colim_{\mathbf{K}}F \to colim_{\mathbf{K}}G$ coincides with:

$$\coprod_{\sigma\in K_0} \Psi_\sigma : \coprod_{\sigma\in K_0} F(\sigma) \to \coprod_{\sigma\in L_0} G(\sigma)$$

Since by assumption each Ψ_σ is an (acyclic) cofibration, Proposition 2.6 (1) implies that so is their coproduct.

Let us assume that the lemma is true for those spaces whose dimension is less than n. Let $dim(K) = n$. For simplicity let us assume that K has only one non-degenerate simplex of dimension n, i.e., K fits into a push-out square:

$$\begin{array}{ccc} \partial\Delta[n] & \hookrightarrow & \Delta[n] \\ \downarrow & & \downarrow \sigma \\ N & \hookrightarrow & K \end{array} \qquad \text{where } dim(N) < n.$$

Consider the following commutative diagram:

$$\begin{array}{ccccccccc} colim_{\mathbf{K}}F & = & colim & (& colim_{\mathbf{N}}F & \longleftarrow & colim_{\boldsymbol{\partial\Delta}[n]}F & \longrightarrow & F(\sigma) \ ) \\ \downarrow {\scriptstyle colim_{\mathbf{K}}\Psi} & & & & \downarrow {\scriptstyle colim_{\mathbf{N}}\Psi} & & \downarrow {\scriptstyle colim_{\boldsymbol{\partial\Delta}[n]}\Psi} & & \downarrow {\scriptstyle \Psi_\sigma} \\ colim_{\mathbf{K}}G & = & colim & (& colim_{\mathbf{N}}G & \longleftarrow & colim_{\boldsymbol{\partial\Delta}[n]}G & \longrightarrow & G(\sigma) \ ) \end{array}$$

We apply Proposition 2.6 (2) to show that $colim_{\mathbf{K}}\Psi$ is an (acyclic) cofibration. The pull-back of Ψ along $N \hookrightarrow L$ is an (acyclic) cofibration, and thus by the inductive assumption so is $colim_{\mathbf{N}}\Psi$.

Since the map $\partial\sigma : \partial\Delta[n] \to K$ can send non-degenerate simplices to degenerate ones, the pull-back of Ψ along $\partial\Delta[n] \to K$ is not a cofibration. Therefore we can not apply the inductive assumption directly to argue that $colim_{\boldsymbol{\partial\Delta}[n]}\Psi$ is an (acyclic) cofibration. However, according to Corollary 19.6 (1) this pull-back is an (acyclic) $\partial\sigma$-cofibration. Let us consider the reduction $\partial\Delta[n] \xrightarrow{\partial\sigma_{red}} red(\partial\sigma) \to K$. By Theorem 18.7, the natural transformation $(\partial\sigma_{red})^k\Psi : (\partial\sigma_{red})^kF \to (\partial\sigma_{red})^kG$ in $Fun^b\big(\mathbf{red}(\partial\sigma), \mathcal{M}\big)$ coincides with the pull-back of Ψ along $red(\partial\sigma) \to K$. Since this map is reduced, it follows that $(\partial\sigma_{red})^k\Psi$ is an (acyclic) id-cofibration. The dimension of $red(\partial\sigma)$ is less than n ($\partial\Delta[n] \to red(\partial\sigma)$ is an epimorphism). Thus by the inductive assumption, $colim_{\mathbf{red}(\partial\sigma)}(\partial\sigma_{red})^k\Psi$ is an (acyclic) cofibration. As the left Kan extension process does not modify colimits (see Proposition 9.2), $colim_{\boldsymbol{\partial\Delta}[n]}\Psi$ is also an (acyclic) cofibration.

The cofibrancy assumption on Ψ implies that $M_\Psi(\sigma) \to G(\sigma)$ is an (acyclic) cofibration. The assumptions of Proposition 2.6 (2) are therefore satisfied, and hence $colim_{\mathbf{K}}\Psi$ is an (acyclic) cofibration.

So far we have proven the lemma in the case when K is finite dimensional. If K is infinite dimensional, by considering the skeleton filtration of K and applying Proposition 2.6 (3), Corollary 8.4 (2), and Proposition 20.1, we can conclude that the lemma is also true in this case. □

PROOF OF THEOREM 20.3. Consider the reduction $L \xrightarrow{f_{red}} red(f) \to K$ of the map f. Since $\Psi : F \to G$ is an (acyclic) f-cofibration, Proposition 19.7 asserts that $(f_{red})^k\Psi : (f_{red})^kF \to (f_{red})^kG$ is an (acyclic) id-cofibration in $Fun^b\big(\mathbf{red}(f), \mathcal{M}\big)$.

Thus Lemma 20.4 implies that $colim_{\mathbf{red}(f)}(f_{red})^k\Psi$ is an (acyclic) cofibration in $\mathcal{M}$. As the left Kan extension process does not modify colimits (see Proposition 9.2), it follows that $colim_{\mathbf{L}}\Psi$ is also an (acyclic) cofibration. □

As a first corollary we get Theorem 13.2.

20.5. COROLLARY. *Let $f : L \to K$ be a map and $\Psi : F \to G$ be a natural transformation in $Fun^b(\mathbf{K}, \mathcal{M})$. If Ψ is a cofibration (respectively a weak equivalence and cofibration), then $colim_{\mathbf{L}}\Psi : colim_{\mathbf{L}}F \to colim_{\mathbf{L}}G$ is a cofibration (respectively a weak equivalence and cofibration) in $\mathcal{M}$.*

PROOF. The corollary follows from Theorem 20.3 and the fact that Ψ pulls-back to a cofibration $f^*\Psi$ in $Fun^b_f(\mathbf{L}, \mathcal{M})$ (see Corollary 19.6 (1)). □

20.6. COROLLARY. *Let $f : L \to K$ be a map.*

1. *If $\Psi : F \to G$ is an (acyclic) f-cofibration, then, for any simplex $\sigma \in L$, $\Psi_\sigma : F(\sigma) \to G(\sigma)$ is an (acyclic) cofibration in $\mathcal{M}$.*
2. *If $G : \mathbf{L} \to \mathcal{M}$ is f-cofibrant, then $colim_{\mathbf{L}}G$ is a cofibrant object in $\mathcal{M}$ and, for any simplex $\sigma \in L$, $G(\sigma)$ is cofibrant in $\mathcal{M}$.*
3. *A natural transformation $\Psi : F \to G$ is an acyclic f-cofibration if and only if Ψ is an f-cofibration and, for any simplex $\sigma \in L$, the morphism $\Psi_\sigma : F(\sigma) \to G(\sigma)$ is a weak equivalence in $\mathcal{M}$.*

PROOF OF 1. Let $\sigma : \Delta[n] \to L$ be a simplex. The pull-back of Ψ along σ is an (acyclic) $(f \circ \sigma)$-cofibration. Thus, according to Theorem 20.3, the map $colim_{\mathbf{\Delta}[n]}F = F(\sigma) \xrightarrow{\Psi_\sigma} G(\sigma) = colim_{\mathbf{\Delta}[n]}G$ is an (acyclic) cofibration.

PROOF OF 3. If Ψ is an acyclic f-cofibration in $Fun^b_f(\mathbf{L}, \mathcal{M})$, then in particular it is an f-cofibration. Now part 1 of the corollary shows that, for any simplex $\sigma \in L$, Ψ_σ is a weak equivalence.

Assume that Ψ is an f-cofibration and, for any $\sigma \in L$, Ψ_σ is a weak equivalence. For any simplex $\sigma : \Delta[n] \to L$, consider the following commutative diagram associated with σ:

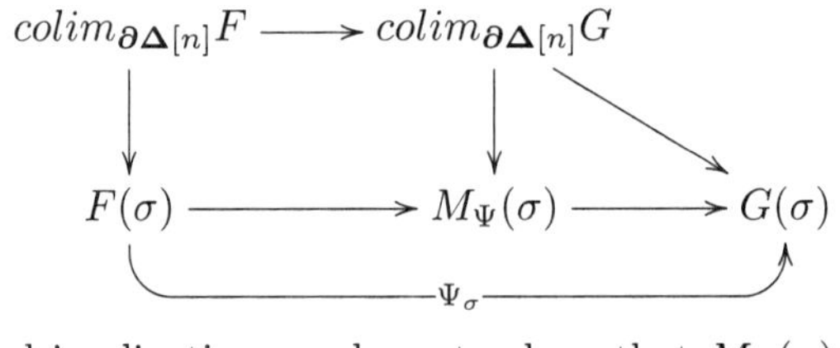

To prove the second implication, we have to show that $M_\Psi(\sigma) \to G(\sigma)$ is a weak equivalence for any $\sigma \in L$. The pull-back of Ψ along $\partial\sigma : \partial\Delta[n] \to L$ is a cofibration relative to $(f \circ \partial\sigma)$. As it is also an objectwise weak equivalence, Theorem 20.3 (2) implies that $colim_{\partial\mathbf{\Delta}[n]}F \to colim_{\partial\mathbf{\Delta}[n]}G$ is an acyclic cofibration. Since Ψ_σ is a weak equivalence, it is clear now that so is $M_\Psi(\sigma) \to G(\sigma)$. □

21. $Fun^b_f(\mathbf{L}, \mathcal{M})$ as a model category

We prove in this section that the category of relatively bounded diagrams, with the choice of cofibrations introduced in Definition 19.1, is a model category. This generalizes the result we stated in the absolute case (Theorem 13.1).

21.1. THEOREM. *Let $f : L \to K$ be a map. The category $Fun^b_f(\mathbf{L}, \mathcal{M})$, together with the following choice of weak equivalences, fibrations, and cofibrations satisfies the axioms of a model category:*

- *a natural transformation $\Psi : F \to G$ is a weak equivalence (respectively a fibration) if for any simplex $\sigma \in L$, $\Psi_\sigma : F(\sigma) \to G(\sigma)$ is a weak equivalence (respectively a fibration) in $\mathcal{M}$;*
- *a natural transformation $\Psi : F \to G$ is a cofibration if it is an f-cofibration in the sense of Definition 19.1.*

21.2. REMARK. Observe that the notions of a weak equivalence, fibration, and cofibration in $Fun^b_f(\mathbf{L}, \mathcal{M})$ are local (see Definition 7.1). A natural transformation $\Psi \in Fun^b_f(\mathbf{L}, \mathcal{M})$ is a weak equivalence, a fibration, or a cofibration if and only if, for any simplex $\sigma : \Delta[n] \to L$, its pull-back $\sigma^*\Psi$ is so in $Fun^b_{f\circ\sigma}(\mathbf{\Delta}[n], \mathcal{M})$.

PROOF OF THEOREM 21.1. Consider the reduction $L \xrightarrow{f_{red}} red(f) \to K$ of f. Checking the axioms of a model category on $Fun^b_f(\mathbf{L}, \mathcal{M})$ can be reduced to checking these axioms for $Fun^b\big(\mathbf{red}(f), \mathcal{M}\big)$ (see Theorem 18.7 and Proposition 19.7), and hence the theorem follows from Theorem 13.1. □

21.3. REMARK. Let $f : L \to K$ be a map and $L \xrightarrow{f_{red}} red(f) \to K$ be its reduction. Observe that the proof of Theorem 21.1 relies on the fact that the pair of adjoint functors:

$$Fun^b_f(\mathbf{L}, \mathcal{M}) \underset{(f_{red})^*}{\overset{(f_{red})^k}{\rightleftarrows}} Fun^b\big(\mathbf{red}(f), \mathcal{M}\big)$$

is a Quillen equivalence (see [**31**]) of model categories.

Using K. Brown's lemma (see Proposition 3.4) and Theorems 21.1 and 20.3, we get:

21.4. COROLLARY. *The colimit functor $colim_{\mathbf{L}} : Fun^b_f(\mathbf{L}, \mathcal{M}) \to \mathcal{M}$ is homotopy meaningful on cofibrant objects.* □

22. Cones

In this section we discuss constructions of cones in the category of spaces and the category of small categories.

A cone over a space K is a space CK which is built by adding an extra vertex to K and joining all the simplices of K with this vertex. Here is the precise definition:

22.1. DEFINITION. Let K be a space. The *cone* over K is the simplicial set CK whose set of n-dimensional simplices is given by:

$$(CK)_n = \{e^{n+1}\} \coprod \Big(\coprod_{i=0}^{n} K_i \times \{e^{n-i}\} \Big)$$

The simplicial operators d_i and s_i are given by:

$$s_0 : (CK)_0 \to (CK)_1 \ , \ s_0(e^1) := e^2 \ , \ s_0(\sigma, e^0) := (s_0\sigma, e^0);$$

for $n > 0$, $dim(\sigma) = i$, and $0 \le j \le n$ the maps $(CK)_{n+1} \stackrel{s_j}{\leftarrow} (CK)_n \stackrel{d_j}{\rightarrow} (CK)_{n-1}$ are defined as:

$$d_j(e^{n+1}) := e^n \ , \ d_j(\sigma, e^{n-i}) := \begin{cases} e^n & \text{if } i = j = 0 \\ \big(d_j(\sigma), e^{n-j}\big) & \text{if } j \le i \text{ and } i > 0 \\ \big(\sigma, e^{n-j-1}\big) & \text{if } j > i \end{cases}$$

$$s_j(e^{n+1}) := e^{n+2} \ , \ s_j(\sigma, e^{n-i}) := \begin{cases} \big(s_j(\sigma), e^{n-i}\big) & \text{if } j \le i \\ \big(\sigma, e^{n+1-i}\big) & \text{if } j > i \end{cases}$$

A simplex (σ, e^k) is non-degenerate in CK if σ is non-degenerate in K and either $k = 0$ or $k = 1$.

The cone construction is natural, i.e., a map $f : L \to K$ induces a map of cones $Cf : CL \to CK$. It sends $e^i \in CL$ and $(\sigma, e^i) \in CL$ to $e^i \in CK$ and $(f(\sigma), e^i) \in CK$ respectively. Observe that there is a natural inclusion $K \hookrightarrow CK$, $K \ni \sigma \mapsto (\sigma, e^0) \in CK$.

Since the colimit commutes with sums, the cone functor commutes with colimits. The map $colim_I CF \to C(colim_I F)$, induced by the natural transformation $C(F \to colim_I F)$, is an isomorphism for any $F : I \to Spaces$. In particular if $K = colim\big(L \leftarrow \partial\Delta[n] \hookrightarrow \Delta[n]\big)$, then $CK = colim\big(CL \leftarrow C\partial\Delta[n] \hookrightarrow C\Delta[n]\big)$. This can be used to build CK inductively on the cell decomposition of K.

22.2. Example. The cone over $C\Delta[n]$ can be identified with $\Delta[n+1]$ where the inclusion $\Delta[n] \hookrightarrow C\Delta[n]$ corresponds to the map $d_{n+1} : \Delta[n] \to \Delta[n+1]$. The cone $C\partial\Delta[n]$ is isomorphic to $\Delta[n+1, n+1]$ and the map $C(\partial\Delta[n] \hookrightarrow \Delta[n])$ corresponds to the inclusion of the $(n+1)$-st horn $\Delta[n+1, n+1] \hookrightarrow \Delta[n+1]$.

22.3. Remark. The opposite category $(\mathbf{CK})^{op}$ has a more transparent interpretation as a Grothendieck construction (see Section 38):

$$(\mathbf{CK})^{op} = Gr\big(\mathbf{K}^{op} \stackrel{pr_1}{\longleftarrow} \mathbf{K}^{op} \times \mathbf{\Delta}[0]^{op} \stackrel{pr_2}{\longrightarrow} \mathbf{\Delta}[0]^{op}\big)$$

where $\mathbf{K}^{op} \times \mathbf{\Delta}[0]^{op}$ denotes the product of $\mathbf{K}^{op}$ and $\mathbf{\Delta}[0]^{op}$ in Cat. We will take advantage of this presentation in Proposition 28.1.

The cone construction in *Spaces* has its analogue in the category of small categories. In Cat the construction is much simpler. To a small category we just add a terminal object.

22.4. Definition. Let I be a small category. The *cone* over I is the category CI defined as follows:

$$ob(CI) = ob(I) \coprod \{e\}$$

$$mor_{CI}(a, b) = \begin{cases} mor_I(a, b) & \text{if } a \ne e, \ b \ne e \\ \{e_a : a \to e\} & \text{if } b = e \\ \emptyset & \text{if } a = e, \ b \ne e \end{cases}$$

The object $e \in CI$ is terminal. As in the case of *Spaces*, the cone construction in Cat is natural, i.e., a functor $f : I \to J$ induces a functor of cones $Cf : CI \to CJ$. Observe that there is a natural inclusion $I \to CI$.

On the level of nerves $N(CI)$ can be identified with $CN(I)$. Under this identification the object $e \in CI$ corresponds to the vertex $e^1 \in CN(I)$ and the map $N(I \to CI)$ corresponds to the natural inclusion $N(I) \hookrightarrow CN(I)$.

23. Diagrams indexed by cones I

In this section we are going to compute ocolimits of certain diagrams indexed by the nerves of categories having a terminal object. This has been already used in the proof of Theorem 15.1 (see Lemma 15.2).

Let K be a space. For any simplex of the form $(\sigma, e^1) \in CK$ there is a unique morphism $e^1 \to (\sigma, e^1)$ corresponding to the inclusion of the vertex e^1 into (σ, e^1).

23.1. PROPOSITION. *Let $F : \mathbf{CK} \to \mathcal{M}$ be cofibrant in $Fun^b(\mathbf{CK}, \mathcal{M})$. If for every simplex of the form $(\sigma, e^1) \in CK$ the morphism $F(e^1) \to F(\sigma, e^1)$ is a weak equivalence, then so is $F(e^1) \to colim_{\mathbf{CK}} F$.*

23.2. LEMMA. *Let $f : L \to K$ be a map. The reduction of $Cf : CL \to CK$ (see Definition 18.6) can be identified with $CL \overset{C(f_{red})}{\longrightarrow} Cred(f) \to CK$.*

PROOF. The lemma follows from the fact that $Cs_i : C\Delta[n+1] \to C\Delta[n]$ can be identified with $s_i : \Delta[n+2] \to \Delta[n+1]$. □

PROOF OF PROPOSITION 23.1. The strategy is the same as in the proof of Lemma 20.4. We first assume that K is finite dimensional. In this case we argue by induction on the dimension of K.

Let $dim(K) = 0$. For simplicity assume $K = \Delta[0]$. The general case can be proven analogously. Since $K = \Delta[0]$, we have $CK = \Delta[1]$. The non-degenerate and 1-dimensional simplex in $\Delta[1]$ corresponds to $(0, e^1) \in CK$. Hence, according to the assumption, $F(e^1) \to colim_{\mathbf{\Delta[1]}} F = F(0, e^1)$ is a weak equivalence.

Let us assume that the proposition holds for those spaces whose dimension is less than n. Let $dim(K) = n$. For simplicity assume that K has only one non-degenerate simplex of dimension n, i.e., K fits into a push-out square:

$$\begin{array}{ccc} \partial\Delta[n] & \hookrightarrow & \Delta[n] \\ \downarrow & & \downarrow \sigma \\ N & \hookrightarrow & K \end{array} \qquad \text{where } dim(N) < n.$$

By applying the cone construction we get another push-out diagram:

$$\begin{array}{ccccc} \Delta[n+1, n+1] = & C\partial\Delta[n] & \hookrightarrow & C\Delta[n] & = \Delta[n+1] \\ & \downarrow & & \downarrow C\sigma & \\ & CN & \hookrightarrow & CK & \end{array}$$

By Corollary 8.4 (1) this induces yet another push-out:

$$\begin{array}{ccc} colim_{\mathbf{C\partial\Delta[n]}} F & \longrightarrow & F(\sigma, e^1) \\ \downarrow & & \downarrow \\ colim_{\mathbf{CN}} F & \longrightarrow & colim_{\mathbf{CK}} F \end{array}$$

Since $CN \hookrightarrow CK$ is a monomorphism, the pull-back $\mathbf{CN} \hookrightarrow \mathbf{CK} \overset{F}{\to} \mathcal{M}$ is a cofibrant diagram. The inductive assumption implies therefore that the map

$F(e^1) \to colim_{\mathbf{CN}}F$ is a weak equivalence. Thus the proposition would be proved once we show that $colim_{\mathbf{C\partial\Delta}[n]}F \to F(\sigma, e^1) \simeq F(e^1)$ is an acyclic cofibration. By Proposition 2.1 this would imply that so is $colim_{\mathbf{CN}}F \to colim_{\mathbf{CK}}F$, and we could conclude that the composite $F(e^1) \to colim_{\mathbf{CN}}F \to colim_{\mathbf{CK}}F$ is a weak equivalence.

The map $C\sigma : C\Delta[n] \to CK$ sends non-degenerate simplices in $C\Delta[n]\setminus C\partial\Delta[n]$ to non-degenerate simplices in CK. Hence cofibrancy of F implies that the morphism $colim_{\mathbf{C\partial\Delta}[n]}F \to F(\sigma, e^1)$ is a cofibration (see Corollary 20.2).

We are going to show that this morphism is also a weak equivalence. As the map $C\partial\sigma : C\partial\Delta[n] \to CK$ can send non-degenerate simplices to degenerate ones, the composite $\mathbf{C\partial\Delta}[n] \to \mathbf{CK} \xrightarrow{F} \mathcal{M}$ is not necessarily a cofibrant diagram. However, this pull-back is $C\partial\sigma$-cofibrant. Let us consider the reduction $C\partial\Delta[n] \xrightarrow{C\partial\sigma_{red}} Cred(\partial\sigma) \to CK$ of $C\partial\sigma$ (see Lemma 23.2). According to Proposition 19.7 the diagram $(C\partial\sigma_{red})^k F$ is cofibrant. Moreover its pull-back along $C\partial\sigma_{red}$ is isomorphic to F, and hence $(C\partial\sigma_{red})^k F$ satisfies the assumption of the proposition. Thus, by the inductive assumption, $F(e^1) \to colim_{\mathbf{Cred}(\partial\sigma)}(C\partial\sigma_{red})^k F$ is a weak equivalence. As the left Kan extension does not modify colimits we can conclude that $F(e^1) \to colim_{\mathbf{C\partial\Delta}[n]}F$ is also a weak equivalence.

So far we have proven the proposition in the case when K is finite dimensional. The infinite dimensional case can be proven by considering the skeleton filtration of K and using Proposition 2.5 (3), Corollary 8.4 (2), and Corollary 20.2. □

23.3. COROLLARY. *Let $F : \mathbf{CK} \to \mathcal{M}$ be a bounded diagram. If for every simplex of the form $(\sigma, e^1) \in CK$, $F(e^1) \to F(\sigma, e^1)$ is a weak equivalence, then $ocolim_{\mathbf{CK}}F$ is weakly equivalent to $F(e^1)$.* □

23.4. COROLLARY. *Let I be a small category and $F : \mathbf{N}(CI) \to \mathcal{M}$ be a bounded and cofibrant diagram. Assume that the diagram F sends any morphism of the form $d_n \circ \cdots \circ d_1 : e \to (i_n \to \cdots \to i_1 \to e)$ in $\mathbf{N}(CI)$ to a weak equivalence in $\mathcal{M}$. Then the morphism $F(e) \to colim_{\mathbf{N}(CI)}F$ is also a weak equivalence.* □

Corollary 23.4 can be generalized to categories with a terminal object.

23.5. PROPOSITION. *Let I be a category with a terminal object denoted by t and $F : \mathbf{N}(I) \to \mathcal{M}$ be a bounded and cofibrant diagram. Assume that F sends any morphism of the form $d_n \circ \cdots \circ d_1 : t \to (i_n \to \cdots \to i_1 \to t)$ in $\mathbf{N}(I)$ to a weak equivalence. Then the morphism $F(t) \to colim_{\mathbf{N}(I)}F$ is also a weak equivalence.*

PROOF. Even though I has a terminal object, in general, $N(I)$ is not isomorphic to a cone. Thus we can not apply Corollary 23.4 directly. However, using the existence of a terminal object $t \in I$ we can construct a retraction $CI \to I$ of $I \hookrightarrow CI$:

$$CI \to I,\ I \ni i \mapsto i,\ e \mapsto t$$

$$I \ni (i \to j) \mapsto (i \to j),\ (e_i : i \to e) \mapsto (i \to t)$$

The composite $I \hookrightarrow CI \to I$ is the identity.

By pulling back F along $N(CI) \to N(I)$ we get a diagram which is no longer cofibrant. Let $QF : \mathbf{N}(CI) \to \mathcal{M}$ together with $QF \overset{\sim}{\twoheadrightarrow} F$ be a cofibrant replacement of the composite $\mathbf{N}(CI) \to \mathbf{N}(I) \xrightarrow{F} \mathcal{M}$ such that the pull-back diagram

$\mathbf{N}(I) \hookrightarrow \mathbf{N}(CI) \overset{QF}{\to} \mathcal{M}$ coincides with $F : \mathbf{N}(I) \to \mathcal{M}$. We can form now the following commutative diagram:

$$\begin{array}{ccccc} F(t) & \xrightarrow{\quad id \quad} & F(t) & \xrightarrow{\quad id \quad} & F(t) \\ \downarrow & & \downarrow & & \downarrow \\ colim_{\mathbf{N}(I)}F & \longrightarrow & colim_{\mathbf{N}(CI)}QF & \longrightarrow & colim_{\mathbf{N}(I)}F \\ \llcorner & \longrightarrow & \underset{id}{\longrightarrow} & \longrightarrow & \nearrow \end{array}$$

The functor $QF : \mathbf{N}(CI) \to \mathcal{M}$ satisfies the assumption of Corollary 23.4, and hence $QF(e) \to colim_{\mathbf{N}(CI)}QF$ is a weak equivalence. It follows that so is the map $F(t) \to colim_{\mathbf{N}(CI)}QF$. According to axiom **MC3** for the model structure on $\mathcal{M}$ the morphism $F(t) \to colim_{\mathbf{N}(I)}F$ is also a weak equivalence. This proves the proposition. $\square$

Lemma 15.2, which was used in the proof of Theorem 15.1, is now a direct consequence. Recall that $\epsilon : \mathbf{N}(I) \to I$ denotes the forgetful functor (see Definition 6.6).

23.6. COROLLARY. *Let I be a small category with a terminal object denoted by t. Let $F : \mathbf{N}(I) \to \mathcal{M}$ be a bounded and cofibrant diagram. Assume that there exists $F' : I \to \mathcal{M}$ and a weak equivalence $F \overset{\sim}{\to} \epsilon^* F'$ in $Fun^b\big(\mathbf{N}(I), \mathcal{M}\big)$. Then the morphism*

$$colim_{\mathbf{N}(I)}F \to colim_{\mathbf{N}(I)}\epsilon^* F' = colim_I F' = F'(t)$$

is a weak equivalence in $\mathcal{M}$. $\square$

Since $N(I^{op}) \cong N(I)$, Proposition 23.5 dualizes to categories with an initial object.

23.7. PROPOSITION. *Let I be a category with an initial object denoted by e. Let $F : \mathbf{N}(I) \to \mathcal{M}$ be a bounded and cofibrant diagram. Assume that F sends any morphism of the form $d_{n-1} \circ \cdots \circ d_0 : e \to (e \to i_{n-1} \to \cdots \to i_0)$ in $\mathbf{N}(I)$ to a weak equivalence. Then the morphism $F(e) \to colim_{\mathbf{N}(I)}F$ is also a weak equivalence.* $\square$

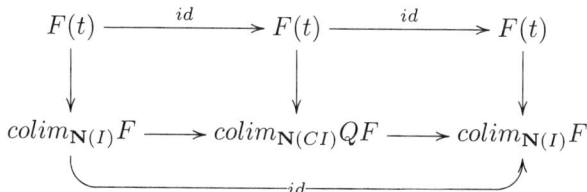

CHAPTER III

Properties of homotopy colimits

24. Fubini theorem

So far we have looked at diagrams indexed by a single space. In Sections 24–26 we consider diagrams indexed by a parameterized collection of spaces. We start with the simplest case: diagrams indexed by the product of two simplex categories. The aim of this section is to prove the so-called Fubini theorem for ocolimits and hocolimits, which asserts that the ocolimit, respectively hocolimit, commutes with itself. It is the homotopy theoretical version of the isomorphisms:

$$colim_{I\times J}F = colim_I colim_J F = colim_J colim_I F$$

Let $\mathcal{M}$ be a model category. Recall that $\mathbf{K}\tilde{\times}\mathbf{N}$ denotes the product of the simplex categories of spaces K and N (see 6.10). Via the standard exponential map $Fun^b\big(\mathbf{K}, Fun^b(\mathbf{N},\mathcal{M})\big)$ can be embedded, as a full subcategory, into $Fun(\mathbf{K}\tilde{\times}\mathbf{N},\mathcal{M})$. Its objects can be characterized as those functors $F:\mathbf{K}\tilde{\times}\mathbf{N}\to\mathcal{M}$ where, for any degeneracy morphisms $s_i\sigma\to\sigma$ in $\mathbf{K}$ and $s_j\tau\to\tau$ in $\mathbf{N}$, the induced natural transformations $F(s_i\sigma,-)\to F(\sigma,-)$ and $F(-,s_i\tau)\to F(-,\tau)$ are isomorphisms. By symmetry we can identify $Fun^b\big(\mathbf{K},Fun^b(\mathbf{N},\mathcal{M})\big)$ with $Fun^b\big(\mathbf{N},Fun^b(\mathbf{K},\mathcal{M})\big)$. This full subcategory of $Fun(\mathbf{K}\tilde{\times}\mathbf{N},\mathcal{M})$ is denoted by $Fun^b(\mathbf{K}\tilde{\times}\mathbf{N},\mathcal{M})$. Its objects are called bounded diagrams.

Bounded diagrams with values in a model category form a model category (see Theorem 13.1). The identifications:

$$Fun^b(\mathbf{K}\tilde{\times}\mathbf{N},\mathcal{M}) = Fun^b\big(\mathbf{K},Fun^b(\mathbf{N},\mathcal{M})\big)$$

$$Fun^b(\mathbf{K}\tilde{\times}\mathbf{N},\mathcal{M}) = Fun^b\big(\mathbf{N},Fun^b(\mathbf{K},\mathcal{M})\big)$$

can be therefore used to induce two model structures on $Fun^b(\mathbf{K}\tilde{\times}\mathbf{N},\mathcal{M})$. We leave to the reader to verify:

24.1. PROPOSITION. *The two model structures on $Fun^b(\mathbf{K}\tilde{\times}\mathbf{N},\mathcal{M})$ coincide.* □

Weak equivalences and fibrations in $Fun^b(\mathbf{K}\tilde{\times}\mathbf{N},\mathcal{M})$ are easy to describe. They are simply objectwise weak equivalences and fibrations. Cofibrations however are more subtle. The following two propositions describe their crucial properties.

24.2. PROPOSITION. *Let $F:\mathbf{K}\tilde{\times}\mathbf{N}\to\mathcal{M}$ be a bounded and cofibrant diagram.*

1. *For all simplices $\sigma\in K$ and $\tau\in N$, the diagrams $F(\sigma,-):\mathbf{N}\to\mathcal{M}$ and $F(-,\tau):\mathbf{K}\to\mathcal{M}$ are bounded and cofibrant.*
2. *The diagrams $colim_{\sigma\in\mathbf{K}}F(\sigma,-):\mathbf{N}\to\mathcal{M}$ and $colim_{\tau\in\mathbf{N}}F(-,\tau):\mathbf{K}\to\mathcal{M}$ are bounded and cofibrant.*
3. *The object $colim_{\mathbf{K}\tilde{\times}\mathbf{N}}F$ is cofibrant in $\mathcal{M}$.*

PROOF. Cofibrancy of the diagram $F : \mathbf{K} \to Fun^b(\mathbf{N}, \mathcal{M})$ means that, for any non-degenerate simplex $\gamma : \Delta[n] \to K$, the natural transformation:

$$colim_{\xi \in \boldsymbol{\partial\Delta}[n]} F(\xi, -) \to F(\gamma, -)$$

is a cofibration in $Fun^b(\mathbf{N}, \mathcal{M})$. We can therefore apply Corollary 20.6 (1) and Theorem 13.2 to conclude that, for any $\tau \in N$, the morphisms:

$$colim_{\xi \in \boldsymbol{\partial\Delta}[n]} F(\xi, \tau) \to F(\gamma, \tau) \qquad colim_{\tau \in \mathbf{N}}\big(colim_{\xi \in \boldsymbol{\partial\Delta}[n]} F(\xi, \tau) \to F(\gamma, \tau)\big)$$

are cofibrations in $\mathcal{M}$. This means exactly that the diagrams:

$$F(-, \tau) : \mathbf{K} \to \mathcal{M} \qquad colim_{\tau \in \mathbf{N}} F(-, \tau) : \mathbf{K} \to \mathcal{M}$$

are cofibrant in $Fun^b(\mathbf{K}, \mathcal{M})$.

By symmetry we get that $F(\sigma, -) : \mathbf{N} \to \mathcal{M}$ and $colim_{\sigma \in \mathbf{K}} F(\sigma, -) : \mathbf{N} \to \mathcal{M}$ are also cofibrant in $Fun^b(\mathbf{N}, \mathcal{M})$.

Part 3 of the proposition follows from part 2 and Corollary 13.4. □

24.3. PROPOSITION. *The functor $colim_{\mathbf{K}\tilde{\times}\mathbf{N}} : Fun^b(\mathbf{K}\tilde{\times}\mathbf{N}, \mathcal{M}) \to \mathcal{M}$ is homotopy meaningful on cofibrant objects.*

PROOF. Let $F : \mathbf{K}\tilde{\times}\mathbf{N} \to \mathcal{M}$, $G : \mathbf{K}\tilde{\times}\mathbf{N} \to \mathcal{M}$ be bounded and cofibrant diagrams and $\Phi : F \xrightarrow{\sim} G$ be a weak equivalence. Proposition 24.2 (1) asserts that, for any $\sigma \in K$, $F(\sigma, -)$ and $G(\sigma, -)$ are cofibrant objects in $Fun^b(\mathbf{N}, \mathcal{M})$. Therefore after applying the colimit we get a weak equivalence in $\mathcal{M}$ (see Corollary 13.4):

$$colim_{\mathbf{N}} \Phi_{(\sigma,-)} : colim_{\mathbf{N}} F(\sigma, -) \xrightarrow{\sim} colim_{\mathbf{N}} G(\sigma, -)$$

The diagrams $\mathbf{K} \ni \sigma \mapsto colim_{\mathbf{N}} F(\sigma, -) \in \mathcal{M}$ and $\mathbf{K} \ni \sigma \mapsto colim_{\mathbf{N}} G(\sigma, -) \in \mathcal{M}$ are also cofibrant objects in $Fun^b(\mathbf{K}, \mathcal{M})$ (by Proposition 24.2 (2)). Thus, by the same argument, $colim_{\mathbf{K}} colim_{\mathbf{N}} \Phi$ is a weak equivalence as well. Since $colim_{\mathbf{K}\tilde{\times}\mathbf{N}} \Phi$ coincides with $colim_{\mathbf{K}} colim_{\mathbf{N}} \Phi$, the proposition is proven. □

24.4. DEFINITION. We denote by $ocolim_{\mathbf{K}\tilde{\times}\mathbf{N}} : Fun^b(\mathbf{K}\tilde{\times}\mathbf{N}, \mathcal{M}) \to Ho(\mathcal{M})$ the total left derived functor of $colim_{\mathbf{K}\tilde{\times}\mathbf{N}} : Fun^b(\mathbf{K}\tilde{\times}\mathbf{N}, \mathcal{M}) \to \mathcal{M}$ (cf. Definition 14.1).

Proposition 24.3 shows that the introduced model structure on $Fun^b(\mathbf{K}\tilde{\times}\mathbf{N}, \mathcal{M})$ can be used to construct $ocolim_{\mathbf{K}\tilde{\times}\mathbf{N}}$ (see Proposition 3.5).

24.5. COROLLARY. *The functor $ocolim_{\mathbf{K}\tilde{\times}\mathbf{N}} : Fun^b(\mathbf{K}\tilde{\times}\mathbf{N}, \mathcal{M}) \to Ho(\mathcal{M})$ exists. It can be constructed by choosing a cofibrant replacement Q in $Fun^b(\mathbf{K}\tilde{\times}\mathbf{N}, \mathcal{M})$ and assigning to $F \in Fun^b(\mathbf{K}\tilde{\times}\mathbf{N}, \mathcal{M})$ the object $colim_{\mathbf{K}\tilde{\times}\mathbf{N}} QF \in Ho(\mathcal{M})$.* □

In the case when $\mathcal{M}$ has a functorial factorization of morphisms into cofibrations followed by acyclic fibrations one can construct a functorial cofibrant replacement $Q : Fun^b(\mathbf{K}\tilde{\times}\mathbf{N}, \mathcal{M}) \to Fun^b(\mathbf{K}\tilde{\times}\mathbf{N}, \mathcal{M})$. This can be then used to construct a rigid ocolimit by taking $colim_{\mathbf{K}\tilde{\times}\mathbf{N}} Q(-) : Fun^b(\mathbf{K}\tilde{\times}\mathbf{N}, \mathcal{M}) \to \mathcal{M}$ (compare with Remarks 14.4 and 16.3).

The following proposition describes one of the crucial global feature of the ocolimit construction, the so-called Fubini theorem.

24.6. PROPOSITION. *Let $\mathcal{M}$ be a model category with a functorial factorization of morphisms into cofibrations followed by acyclic fibrations. Then, for any bounded diagram $F : \mathbf{K}\tilde{\times}\mathbf{N} \to \mathcal{M}$, we have the following weak equivalences in $\mathcal{M}$:*

$$ocolim_{\mathbf{K}\tilde{\times}\mathbf{N}} F \simeq ocolim_{\mathbf{K}} ocolim_{\mathbf{N}} F \simeq ocolim_{\mathbf{N}} ocolim_{\mathbf{K}} F$$

PROOF. Let $Q : Fun^b(\mathbf{K}\tilde{\times}\mathbf{N}, \mathcal{M}) \to Fun^b(\mathbf{K}\tilde{\times}\mathbf{N}, \mathcal{M})$ be a functorial cofibrant replacement. Proposition 24.2 (1) implies that, for any $\sigma \in K$ and $\tau \in N$, the diagrams $QF(\sigma, -) : \mathbf{N} \to \mathcal{M}$ and $QF(-, \tau) : \mathbf{K} \to \mathcal{M}$ are cofibrant replacements of $F(\sigma, -) : \mathbf{N} \to \mathcal{M}$ and $F(-, \tau) : \mathbf{K} \to \mathcal{M}$ respectively. It follows that:

$$colim_{\mathbf{N}} QF(\sigma, -) \simeq ocolim_{\mathbf{N}} F(\sigma, -) \qquad colim_{\mathbf{K}} QF(-, \tau) \simeq ocolim_{\mathbf{K}} F(-, \tau)$$

Since the diagrams $\mathbf{K} \ni \sigma \mapsto colim_{\mathbf{N}} QF(\sigma, -)$ and $\mathbf{N} \ni \tau \mapsto colim_{\mathbf{K}} QF(-, \tau)$ are also cofibrant (see Proposition 24.2 (2)), we get weak equivalences:

$$colim_{\mathbf{K}} colim_{\mathbf{N}} QF(\sigma, -) \simeq ocolim_{\mathbf{K}} colim_{\mathbf{N}} QF(\sigma, -)$$

$$colim_{\mathbf{N}} colim_{\mathbf{K}} QF(-, \tau) \simeq ocolim_{\mathbf{N}} colim_{\mathbf{K}} QF(-, \tau)$$

The proposition clearly follows. □

Let I and J be small categories and $l : \mathcal{M} \rightleftarrows \mathcal{C} : r$ be a left model approximation. According to Theorem 11.3 (1), one left model approximation of $Fun(I \times J, \mathcal{C})$ is given by the Bousfield-Kan approximation $Fun^b(\mathbf{N}(I \times J), \mathcal{M}) \rightleftarrows Fun^b(I \times J, \mathcal{C})$. In the next theorem we are going to show that $Fun(I \times J, \mathcal{C})$ can also be approximated by $Fun^b(\mathbf{N}(I)\tilde{\times}\mathbf{N}(J), \mathcal{M})$. Let us denote by $\epsilon : \mathbf{N}(I)\tilde{\times}\mathbf{N}(J) \to I \times J$ the product of the forgetful functors $\epsilon : \mathbf{N}(I) \to I$ and $\epsilon : \mathbf{N}(J) \to J$. Recall also that:

$$\epsilon^* : Fun(I \times J, \mathcal{M}) \to Fun^b(\mathbf{N}(I)\tilde{\times}\mathbf{N}(J), \mathcal{M})$$

$$\epsilon^k : Fun^b(\mathbf{N}(I)\tilde{\times}\mathbf{N}(J), \mathcal{M}) \to Fun(I \times J, \mathcal{M})$$

denote respectively the pull-back process and the left Kan extension along ϵ.

24.7. THEOREM. *The pair of adjoint functors:*

$$Fun^b(\mathbf{N}(I)\tilde{\times}\mathbf{N}(J), \mathcal{M}) \underset{\epsilon^* \circ r}{\overset{l \circ \epsilon^k}{\rightleftarrows}} Fun(I \times J, \mathcal{C})$$

is a left model approximation. Moreover if $\mathcal{C}$ is closed under colimits, then this approximation is good for $colim_{I \times J}$.

PROOF. According to Theorem 15.1, the Bousfield-Kan approximation:

$$l : Fun^b(\mathbf{N}(J), \mathcal{M}) \leftrightarrows Fun(J, \mathcal{C}) : r$$

is a left model approximation. We can therefore use the same theorem to conclude that the induced Bousfield-Kan approximation:

$$Fun^b\Big(\mathbf{N}(I), Fun^b(\mathbf{N}(J), \mathcal{M})\Big) \underset{\epsilon^* \circ r}{\overset{l \circ \epsilon^k}{\rightleftarrows}} Fun(I, Fun(I, \mathcal{C}))$$

is also a left model approximation. It is not difficult to see that this pair of adjoint functors can be identified with the one induced by $\epsilon : \mathbf{N}(I)\tilde{\times}\mathbf{N}(J) \to I \times J$. Thus the first part of the theorem clearly follows.

Let us assume that $\mathcal{C}$ is closed under colimits. To prove the second part of the theorem, we need to show that the following composite:

$$Fun^b(\mathbf{N}(I)\tilde{\times}\mathbf{N}(J), \mathcal{M}) \xrightarrow{\epsilon^k} Fun(I \times J, \mathcal{M}) \xrightarrow{l} Fun(I \times J, \mathcal{C}) \xrightarrow{colim_{I \times J}} \mathcal{C}$$

is homotopy meaningful on cofibrant objects. As left adjoints commute with colimits and left Kan extensions do not modify them (see Proposition 36.2 (2)), the above composite coincides with:

$$Fun^b\big(\mathbf{N}(I)\tilde{\times}\mathbf{N}(J),\mathcal{M}\big) \xrightarrow{colim_{\mathbf{N}(I)\tilde{\times}\mathbf{N}(J)}} \mathcal{M} \xrightarrow{l} \mathcal{C}$$

The second part of the theorem is now a consequence of two facts: the functor $colim_{\mathbf{N}(I)\tilde{\times}\mathbf{N}(J)}$ preserves cofibrancy (see Proposition 24.2 (3)) and both of the functors l and $colim_{\mathbf{N}(I)\tilde{\times}\mathbf{N}(J)}$ are homotopy meaningful on cofibrant objects (see Proposition 24.3). □

24.8. COROLLARY. *Let $\mathcal{C}$ be a category closed under colimits and $l : \mathcal{M} \rightleftarrows \mathcal{C} : r$ be a left model approximation. Then the composite:*

$$\begin{array}{ccccc} Fun(I\times J,\mathcal{C}) & \xrightarrow{r} & Fun(I\times J,\mathcal{M}) & \xrightarrow{\epsilon^*} & Fun^b\big(\mathbf{N}(I)\tilde{\times}\mathbf{N}(J),\mathcal{M}\big) \\ & & & & \downarrow{\scriptstyle ocolim_{\mathbf{N}(I)\tilde{\times}\mathbf{N}(J)}} \\ & & Ho(\mathcal{C}) & \xleftarrow{\quad l \quad} & Ho(\mathcal{M}) \end{array}$$

is the total left derived functor of $colim_{I\times J}$. □

As a corollary of Proposition 24.6 and Corollary 24.8 we get the so-called Fubini Theorem for homotopy colimits.

24.9. THEOREM. *Let $\mathcal{C}$ be closed under colimits and $l : \mathcal{M} \leftrightarrows \mathcal{C} : r$ be a left model approximation such that $\mathcal{M}$ has a functorial factorization of morphisms into cofibrations followed by acyclic fibrations. Then for any $F : I\times J \to \mathcal{C}$, we have the following weak equivalences in $\mathcal{C}$:*

$$hocolim_{I\times J}F \simeq hocolim_I hocolim_J F \simeq hocolim_J hocolim_I F$$

□

25. Bounded diagrams indexed by Grothendieck constructions

In Section 24 we introduced the notion of boundedness for diagrams indexed by the product of simplex categories. In this section we generalize this notion to diagrams indexed by Grothendieck constructions.

Let $H : \mathbf{K} \to Spaces$ be a diagram. We can think about its values as simplex categories, i.e., take the composite $\mathbf{H} : \mathbf{K} \to Spaces \to Cat$, and form its Grothendieck construction $Gr_{\mathbf{K}}\mathbf{H}$ (see Definition 38). We would like to understand the local data needed to describe a functor indexed by $Gr_{\mathbf{K}}\mathbf{H}$.

25.1. DEFINITION. Let $H : \mathbf{K} \to Spaces$ be a functor. We say that a family of functors $F = \{F_\sigma : \mathbf{\Delta}[n] \to Fun\big(\mathbf{H}(\sigma),\mathcal{C}\big)\}_{(\sigma:\Delta[n]\to K)}$ is *compatible* over H if, for any simplices $\sigma : \Delta[n] \to K$ and $\xi : \Delta[m] \to K$, and for any morphism $\alpha : \sigma \to \xi$ in $\mathbf{K}$, the following diagram commutes:

$$\begin{array}{ccc} \Delta[n] & \xrightarrow{F_\sigma} & Fun\big(\mathbf{H}(\sigma),\mathcal{C}\big) \\ {\scriptstyle\alpha}\downarrow & & \downarrow{\scriptstyle H(\alpha)^k} \\ \Delta[m] & \xrightarrow{F_\xi} & Fun\big(\mathbf{H}(\xi),\mathcal{C}\big) \end{array}$$

where $H(\alpha)^k$ is the left Kan extension along $H(\alpha) : H(\sigma) \to H(\xi)$.

Let F and G be compatible families of functors over H. A family of natural transformations $\Psi = \{\Psi_\sigma : F_\sigma \to G_\sigma\}_{\sigma \in K}$ is called a morphism from F to G if, for any $\alpha : \sigma \to \xi$ in K, the pull-back $\alpha^* \Psi_\xi$ coincides with $H(\alpha)^k(\Psi_\sigma)$.

Compatible families over H, together with morphisms between them as defined above, clearly form a category. Compatible families in the context of arbitrary small categories (not only for simplex categories) are discussed in more details in Section 40 of Appendix B.

Let ι be the only non-degenerate n-dimensional simplex in $\Delta[n]$ (ι corresponds to the map $id : \Delta[n] \to \Delta[n]$). With any compatible family $F = \{F_\sigma\}_{\sigma \in K}$ over H we can associate a functor $F : Gr_{\mathbf{K}}\mathbf{H} \to \mathcal{C}$. It assigns to $(\sigma, \tau) \in Gr_{\mathbf{K}}\mathbf{H}$ the object $F_\sigma(\iota)(\tau) \in \mathcal{C}$. In this way we get a functor from the category of compatible families over H to $Fun(Gr_{\mathbf{K}}\mathbf{H}, \mathcal{C})$. One can show that this functor is an isomorphism of categories (see Section 40). Thus we do not distinguish between diagrams indexed by $Gr_{\mathbf{K}}\mathbf{H}$ and compatible families over H. We also use the same symbol $Fun(Gr_{\mathbf{K}}\mathbf{H}, \mathcal{C})$ to denote both the category of compatible family of functors over H and the category of functors indexed by $Gr_{\mathbf{K}}\mathbf{H}$.

25.2. Remark. Let $N : \mathbf{K} \to Spaces$ be the constant functor with value N. The Grothendieck construction $Gr_{\mathbf{K}}\mathbf{N}$ can be identified with the product of the simplex categories $\mathbf{K}\tilde{\times}\mathbf{N}$ (see Example 38.1). The colimit functor of diagrams indexed by $\mathbf{K}\tilde{\times}\mathbf{N}$ is often studied using the symmetry of this product. The key property of $colim_{\mathbf{K}\tilde{\times}\mathbf{N}}$ is given by natural isomorphisms:

$$colim_{\mathbf{K}\tilde{\times}\mathbf{N}} F = colim_{\mathbf{K}} colim_{\mathbf{N}} F = colim_{\mathbf{N}} colim_{\mathbf{K}} F$$

For an arbitrary diagram $H : \mathbf{K} \to Spaces$ such a symmetry does not hold. However locally we still can exchange the colimit operation. Let $F : Gr_{\mathbf{K}}\mathbf{H} \to \mathcal{C}$ be a diagram and $\{F_\sigma : \Delta[n] \to Fun(\mathbf{H}(\sigma), \mathcal{C})\}$ be the associated compatible family over H. Then for any simplex $\sigma : \Delta[n] \to K$, there is a natural isomorphism:

$$colim_{\mathbf{H}(\sigma)} colim_{\boldsymbol{\partial\Delta}[n]} F_\sigma = colim_{\sigma' \in \boldsymbol{\partial\Delta}[n]} colim_{\mathbf{H}(\sigma')} F$$

A compatible family over H carries a lot of redundant data just to describe a functor indexed by $Gr_{\mathbf{K}}\mathbf{H}$. However, this way of thinking about diagrams over $Gr_{\mathbf{K}}\mathbf{H}$ helps in keeping track of their local data. This local information is important to describe a notion of boundedness for diagrams indexed by $Gr_{\mathbf{K}}\mathbf{H}$, as well as a model structure on such diagrams.

25.3. Definition. Let $H : \mathbf{K} \to Spaces$ be a bounded diagram. A functor $F : Gr_{\mathbf{K}}\mathbf{H} \to \mathcal{C}$ is called *bounded* if, for any simplex $\sigma : \Delta[n] \to K$,

- $F_\sigma : \boldsymbol{\Delta}[n] \to Fun(\mathbf{H}(\sigma), \mathcal{C})$ has bounded values;
- $F_\sigma : \boldsymbol{\Delta}[n] \to Fun^b(\mathbf{H}(\sigma), \mathcal{C})$ is σ-bounded (see Definition 17.1).

The full subcategory of $Fun(Gr_{\mathbf{K}}\mathbf{H}, \mathcal{C})$ consisting of bounded diagrams is denoted by $Fun^b(Gr_{\mathbf{K}}\mathbf{H}, \mathcal{C})$.

An object $(\sigma, \tau) \in Gr_{\mathbf{K}}\mathbf{H}$ is called non-degenerate if σ is non-degenerate in K and τ is non-degenerate in $H(\sigma)$. A bounded diagram $F : Gr_{\mathbf{K}}\mathbf{H} \to \mathcal{C}$ is determined, up to isomorphism, by the values it takes on the non-degenerate objects in $Gr_{\mathbf{K}}\mathbf{H}$.

When $N : \mathbf{K} \to Spaces$ is the constant diagram with value N the boundedness condition for a diagram indexed by $Gr_{\mathbf{K}}\mathbf{N} = \mathbf{K}\tilde{\times}\mathbf{N}$ coincides with the one given in Section 24.

In the case of a constant diagram $N : \mathbf{K} \to Spaces$ we used the symmetry of the product $Gr_{\mathbf{K}}\mathbf{N} = \mathbf{K}\tilde{\times}\mathbf{N} = Gr_{\mathbf{N}}\mathbf{K}$ to impose a model structure on $Fun^b(Gr_{\mathbf{K}}\mathbf{N}, \mathcal{C})$ (see Section 24). For a general bounded diagram $H : \mathbf{K} \to Spaces$ such a symmetry does not hold. To find an appropriate model structure on $Fun^b(Gr_{\mathbf{K}}\mathbf{H}, \mathcal{C})$ we need to use other methods. In this case the crucial role is played by the local data. As in the case of boundedness (see Definition 25.3) definitions of weak equivalences, fibrations, and cofibrations in $Fun^b(Gr_{\mathbf{K}}\mathbf{H}, \mathcal{C})$ have a local nature.

25.4. Definition. Let $H : \mathbf{K} \to Spaces$ be a bounded diagram, $\mathcal{M}$ be a model category, and $\Psi : F \to G$ be a natural transformation in $Fun^b(Gr_{\mathbf{K}}\mathbf{H}, \mathcal{M})$. We call Ψ a weak equivalence, fibration, or cofibration if, for any $\sigma : \Delta[n] \to K$, $\Psi_\sigma : F_\sigma \to G_\sigma$ is respectively a weak equivalence, fibration, or cofibration in $Fun^b_\sigma(\mathbf{\Delta}[n], Fun^b(\mathbf{H}(\sigma), \mathcal{M}))$.

Using exactly the same arguments as in the proof of Theorem 13.1 one can show:

25.5. Theorem. *Let $H : \mathbf{K} \to Spaces$ be a bounded diagram and $\mathcal{M}$ be a model category. The category $Fun^b(Gr_{\mathbf{K}}\mathbf{H}, \mathcal{M})$, together with the choice of weak equivalences, fibrations, and cofibrations given in Definition 25.4, satisfies the axioms of a model category.* □

Weak equivalences and fibrations in $Fun^b(Gr_{\mathbf{K}}\mathbf{H}, \mathcal{M})$ are easy to describe in terms of diagrams indexed by $Gr_{\mathbf{K}}\mathbf{H}$. They are simply objectwise weak equivalences and fibrations. The analogous description for cofibrations is more complicated. For example, a bounded diagram $F : Gr_{\mathbf{K}}\mathbf{H} \to \mathcal{C}$ is cofibrant if and only if, for any non-degenerate simplex $\sigma : \Delta[n] \to K$, the natural transformation $colim_{\tau \in \partial\mathbf{\Delta}[n]} F(\tau, -) \to F(\sigma, -)$ is a cofibration in $Fun^b(\mathbf{H}(\sigma), \mathcal{M})$.

In the case of the constant diagram $N : \mathbf{K} \to Spaces$ with value N the model structure on $Fun^b(Gr_{\mathbf{K}}\mathbf{N}, \mathcal{M}) = Fun^b(\mathbf{K}\tilde{\times}\mathbf{N}, \mathcal{M})$ given in Theorem 25.5 coincides with the one given in Section 24.

26. Thomason's theorem

In this section we prove the so-called Thomason Theorem ([**47**]). It is the homotopy theoretical version of the isomorphism (see Proposition 40.2):

$$colim_{Gr_I H} F \cong colim_{i \in I} colim_{H(i)} F$$

The strategy is the same as in the proof of our Fubini theorems in Section 24 (see Proposition 24.6 and Theorem 24.9). The following propositions are generalizations of Propositions 24.2 and 24.3. Let $H : \mathbf{K} \to Spaces$ be a bounded diagram and $\mathcal{M}$ be a model category.

26.1. Proposition. *Let $F : Gr_{\mathbf{K}}\mathbf{H} \to \mathcal{M}$ be a bounded and cofibrant diagram.*

1. *For all $\sigma : \Delta[n] \to K$, the diagram $F(\sigma, -) : \mathbf{H}(\sigma) \to \mathcal{M}$ is bounded and cofibrant.*
2. *The diagram $colim_{\mathbf{H}(-)} F : \mathbf{K} \to \mathcal{M}$, $\sigma \mapsto colim_{\mathbf{H}(\sigma)} F(\sigma, -)$, is bounded and cofibrant.*
3. *The object $colim_{Gr_{\mathbf{K}}\mathbf{H}} F$ is cofibrant in $\mathcal{M}$.*

Proof of 1. Cofibrancy of F means that, for any $\sigma : \Delta[n] \to K$, the diagram $F_\sigma : \mathbf{\Delta}[n] \to Fun^b(\mathbf{H}(\sigma), \mathcal{M})$ is σ-cofibrant. It follows that for any $\tau \in \Delta[n]$, $F_\sigma(\tau)$ is a cofibrant object in $Fun^b(\mathbf{H}(\sigma), \mathcal{M})$ (see Corollary 20.6 (2)). Let $\iota \in \Delta[n]$ be the non-degenerate simplex of dimension n. Since $F_\sigma(\iota) = F(\sigma, -)$ the first part of the proposition is proven.

PROOF OF 2. The boundedness of $colim_{\mathbf{H}(-)}F$ is clear. To check its cofibrancy we have to show that, for any non-degenerate simplex $\sigma : \Delta[n] \to K$, the morphism $colim_{\sigma' \in \partial\Delta[n]} colim_{\mathbf{H}(\sigma')}F \to colim_{\mathbf{H}(\sigma)}F$ is a cofibration in $\mathcal{M}$.

Consider $F_\sigma : \Delta[n] \to Fun^b(\mathbf{H}(\sigma), \mathcal{M})$. Let $\iota \in (\Delta[n])_n$ be the non-degenerate simplex. Since $\sigma : \Delta[n] \to K$ is non-degenerate and F_σ is σ-cofibrant, by definition, $colim_{\partial\Delta[n]}F_\sigma \to F_\sigma(\iota)$ is a cofibration in $Fun^b(\mathbf{H}(\sigma), \mathcal{M})$. Thus we can apply Lemma 20.4 to conclude that $colim_{\mathbf{H}(\sigma)} colim_{\partial\Delta[n]}F_\sigma \to colim_{\mathbf{H}(\sigma)}F_\sigma(\iota)$ is a cofibration in $\mathcal{M}$. The second part of the proposition now follows since this morphism can be identified with $colim_{\sigma' \in \partial\Delta[n]} colim_{\mathbf{H}(\sigma')}F \to colim_{\mathbf{H}(\sigma)}F$ (see Remark 25.2).

PROOF OF 3. The third part is a consequence of part 2, Corollary 13.4, and the fact that $colim_{Gr_{\mathbf{K}}\mathbf{H}}F = colim_{\sigma \in \mathbf{K}} colim_{\mathbf{H}(\sigma)}F$ (see Proposition 40.2). □

26.2. PROPOSITION. *The functor $colim_{Gr_{\mathbf{K}}\mathbf{H}} : Fun^b(Gr_{\mathbf{K}}\mathbf{H}, \mathcal{M}) \to \mathcal{M}$ is homotopy meaningful on cofibrant objects.*

PROOF. Let $F : Gr_{\mathbf{K}}\mathbf{H} \to \mathcal{M}$, $G : Gr_{\mathbf{K}}\mathbf{H} \to \mathcal{M}$ be bounded and cofibrant diagrams and $\Psi : F \xrightarrow{\sim} G$ be a weak equivalence. Proposition 26.1 (1) asserts that, for any $\sigma \in K$, $F(\sigma, -)$ and $G(\sigma, -)$ are cofibrant objects in $Fun^b(\mathbf{H}(\sigma), \mathcal{M})$. Therefore after applying the colimit we get a weak equivalence (see Corollary 13.4):

$$colim_{\mathbf{H}(\sigma)}\Psi_{(\sigma,-)} : colim_{\mathbf{H}(\sigma)}F(\sigma, -) \xrightarrow{\sim} colim_{\mathbf{H}(\sigma)}G(\sigma, -)$$

As the diagrams $colim_{\mathbf{H}(-)}F : \mathbf{K} \to \mathcal{M}$ and $colim_{\mathbf{H}(-)}G : \mathbf{K} \to \mathcal{M}$ are also cofibrant (see Proposition 26.1 (2)), $colim_{\mathbf{K}} colim_{\mathbf{H}(-)}\Psi$ is a weak equivalence as well. Since the morphism $colim_{Gr_{\mathbf{K}}\mathbf{H}}\Psi$ coincides with $colim_{\mathbf{K}} colim_{\mathbf{H}(-)}\Psi$, the proposition is proven. □

26.3. DEFINITION. We denote by $ocolim_{Gr_{\mathbf{K}}\mathbf{H}} : Fun^b(Gr_{\mathbf{K}}\mathbf{H}, \mathcal{M}) \to Ho(\mathcal{M})$ the total left derived functor of $colim_{Gr_{\mathbf{K}}\mathbf{H}} : Fun^b(Gr_{\mathbf{K}}\mathbf{H}, \mathcal{M}) \to \mathcal{M}$.

Proposition 26.2 shows that the introduced model structure on the category $Fun^b(Gr_{\mathbf{K}}\mathbf{H}, \mathcal{M})$ can be used to construct $ocolim_{Gr_{\mathbf{K}}\mathbf{H}}$ (see Proposition 3.5).

26.4. COROLLARY. *The functor $ocolim_{Gr_{\mathbf{K}}\mathbf{H}}$ exists. It can be constructed by choosing a cofibrant replacement Q in $Fun^b(Gr_{\mathbf{K}}\mathbf{H}, \mathcal{M})$ and assigning to a diagram $F \in Fun^b(Gr_{\mathbf{K}}\mathbf{H}, \mathcal{M})$ the object $colim_{Gr_{\mathbf{K}}\mathbf{H}}QF \in Ho(\mathcal{M})$.* □

In the case when $\mathcal{M}$ has a functorial factorization of morphisms into cofibrations followed by acyclic fibrations one can construct a functorial cofibrant replacement $Q : Fun^b(Gr_{\mathbf{K}}\mathbf{H}, \mathcal{M}) \to Fun^b(Gr_{\mathbf{K}}\mathbf{H}, \mathcal{M})$. This can be then used to construct a rigid ocolimit by taking $colim_{Gr_{\mathbf{K}}\mathbf{H}}Q(-) : Fun^b(Gr_{\mathbf{K}}\mathbf{H}, \mathcal{M}) \to \mathcal{M}$ (compare with Remarks 14.4 and 16.3).

We are now ready to prove Thomason's theorem for ocolimits.

26.5. PROPOSITION. *Assume that $\mathcal{M}$ is a model category with a functorial factorization of morphisms into cofibrations followed by acyclic fibrations. Then, for any $F \in Fun^b(Gr_{\mathbf{K}}\mathbf{H}, \mathcal{M})$, we have a weak equivalence in $\mathcal{M}$:*

$$ocolim_{Gr_{\mathbf{K}}\mathbf{H}}F \simeq ocolim_{\sigma \in \mathbf{K}} ocolim_{\mathbf{H}(\sigma)}F$$

PROOF. Let $Q : Fun^b(Gr_{\mathbf{K}}\mathbf{H}, \mathcal{M}) \to Fun^b(Gr_{\mathbf{K}}\mathbf{H}, \mathcal{M})$ be a functorial cofibrant replacement. Proposition 26.1 (1) asserts that for any $\sigma \in K$, the diagram $QF(\sigma, -) : \mathbf{H}(\sigma) \to \mathcal{M}$ is a cofibrant replacement of $F(\sigma, -) : \mathbf{H}(\sigma) \to \mathcal{M}$. It follows that:

$$colim_{\mathbf{H}(\sigma)} QF(\sigma, -) \simeq ocolim_{\mathbf{H}(\sigma)} F(\sigma, -)$$

Since the diagram $colim_{\mathbf{H}(-)} QF : \mathbf{K} \to \mathcal{M}$, $\sigma \mapsto colim_{\mathbf{H}(\sigma)} QF(\sigma, -)$ is also cofibrant (see Proposition 26.1 (2)), we get a weak equivalence:

$$colim_{\sigma \in \mathbf{K}} colim_{\mathbf{H}(\sigma)} QF \simeq ocolim_{\sigma \in \mathbf{K}} colim_{\mathbf{H}(\sigma)} QF$$

The proposition clearly follows. □

Let $l : \mathcal{M} \rightleftarrows \mathcal{C} : r$ be a left model approximation and $H : I \to Cat$ be a functor. Consider the forgetful functors $\epsilon : \mathbf{N}(I) \to I$ and $\epsilon : \mathbf{N}\big(H(i)\big) \to H(i)$. We use the same symbol $\epsilon : Gr_{\mathbf{N}(I)}\mathbf{N}(H) \to Gr_I H$ to denote the induced functor on the level of Grothendieck constructions. Since the above forgetful functors are cofinal with respect to taking colimits, then so is $\epsilon : Gr_{\mathbf{N}(I)}\mathbf{N}(H) \to Gr_I H$. Recall that:

$$\epsilon^* : Fun(Gr_I H, \mathcal{M}) \to Fun^b\big(Gr_{\mathbf{N}(I)}\mathbf{N}(H), \mathcal{M}\big)$$

$$\epsilon^k : Fun^b\big(Gr_{\mathbf{N}(I)}\mathbf{N}(H), \mathcal{M}\big) \to Fun(Gr_I H, \mathcal{M})$$

denote respectively the pull-back process and the left Kan extension along ϵ.

We are going to use $\epsilon : Gr_{\mathbf{N}(I)}\mathbf{N}(H) \to Gr_I H$ to approximate the category $Fun(Gr_I H, \mathcal{C})$ by the model category $Fun^b\big(Gr_{\mathbf{N}(I)}\mathbf{N}(H), \mathcal{M}\big)$. In this way we get a convenient construction of $hocolim_{Gr_I H}$.

26.6. THEOREM. *The pair of adjoint functors:*

$$Fun^b\big(Gr_{\mathbf{N}(I)}\mathbf{N}(H), \mathcal{M}\big) \underset{\epsilon^* \circ r}{\overset{l \circ \epsilon^k}{\rightleftarrows}} Fun(Gr_I H, \mathcal{C})$$

is a left model approximation. Moreover if $\mathcal{C}$ is closed under colimits, then this approximation is good for $colim_{Gr_I H}$.

PROOF. Conditions 1 and 2 of Definition 5.1 are clearly satisfied.

To show that condition 3 is satisfied we need to prove that the composite:

$$Fun^b\big(Gr_{\mathbf{N}(I)}\mathbf{N}(H), \mathcal{M}\big) \xrightarrow{\epsilon^k} Fun(Gr_I H, \mathcal{M}) \xrightarrow{l} Fun(Gr_I H, \mathcal{C})$$

is homotopy meaningful on cofibrant objects. Let F and G be cofibrant diagrams in $Fun^b\big(Gr_{\mathbf{N}(I)}\mathbf{N}(H), \mathcal{M}\big)$ and $\Psi : F \xrightarrow{\sim} G$ be a weak equivalence. By definition $\epsilon^k\Psi$ assigns to $(i, x) \in Gr_I H$ the following morphism in $\mathcal{M}$ (see Section 36):

$$colim_{\epsilon\downarrow(i,x)}\Psi : colim_{\epsilon\downarrow(i,x)} F \to colim_{\epsilon\downarrow(i,x)} G$$

The category $\epsilon\downarrow(i, x)$ can be identified with $Gr_{\mathbf{N}(I\downarrow i)}\mathbf{N}(H\downarrow x)$ and the functor $\epsilon\downarrow(i, x) \to Gr_{\mathbf{N}(I)}\mathbf{N}(H)$ with $Gr_{\mathbf{N}(I\downarrow i)}\mathbf{N}(H\downarrow x) \to Gr_{\mathbf{N}(I)}\mathbf{N}(H)$, which is induced by the maps $N(I\downarrow i) \to N(I)$ and $N\big(H(i)\downarrow x\big) \to N\big(H(i)\big)$ (see Section 39). Since these maps are reduced, the composites:

$$Gr_{\mathbf{N}(I\downarrow i)}\mathbf{N}(H\downarrow x) \to Gr_{\mathbf{N}(I)}\mathbf{N}(H) \xrightarrow{F} \mathcal{M}$$

$$Gr_{\mathbf{N}(I\downarrow i)}\mathbf{N}(H\downarrow x) \to Gr_{\mathbf{N}(I)}\mathbf{N}(H) \xrightarrow{G} \mathcal{M}$$

are cofibrant diagrams. Thus Proposition 26.2 asserts that:

$$(\epsilon^k\Psi)_{(i,x)} = colim_{Gr_{\mathbf{N}(I\downarrow i)}\mathbf{N}(H\downarrow x)}\Psi$$

is a weak equivalence in $\mathcal{M}$. As the objects $(\epsilon^k F)(i,x) = colim_{Gr_{\mathbf{N}(I\downarrow i)}\mathbf{N}(H\downarrow x)}F$ and $(\epsilon^k G)(i,x) = colim_{Gr_{\mathbf{N}(I\downarrow i)}\mathbf{N}(H\downarrow x)}G$ are also cofibrant in $\mathcal{M}$ (see Proposition 26.1 (3)), and l is homotopy meaningful on cofibrant objects, we have that $l(\epsilon^k\Psi)$ is a weak equivalence in $Fun(Gr_I H, \mathcal{C})$.

Consider the composite:

$$Fun^b\big(Gr_{\mathbf{N}(I)}\mathbf{N}(H), \mathcal{M}\big) \xleftarrow{\epsilon^*} Fun(Gr_I H, \mathcal{M}) \xleftarrow{r} Fun(Gr_I H, \mathcal{C})$$

To show that condition 4 of Definition 5.1 is satisfied we need to check that for any $F' : Gr_I H \to \mathcal{C}$, if $F : Gr_{\mathbf{N}(I)}\mathbf{N}(H) \to \mathcal{M}$ is bounded, cofibrant, and $F \to \epsilon^* rF'$ is a weak equivalence, then so is its adjoint $l\epsilon^k F \to F'$. We first show that $\epsilon^k F \to rF'$ is a weak equivalence. To do so we compare these diagrams with a third one.

As in the proof of condition 3, for any $(i,x) \in Gr_I H$, the composite:

$$\epsilon\downarrow(i,x) = Gr_{\mathbf{N}(I\downarrow i)}\mathbf{N}(H\downarrow x) \to Gr_{\mathbf{N}(I)}\mathbf{N}(H) \xrightarrow{F} \mathcal{M}$$

is a cofibrant diagram. It follows that $colim_{\mathbf{N}(H\downarrow x)(-)}F : \mathbf{N}(I\downarrow i) \to \mathcal{M}$ is a cofibrant object in $Fun^b\big(\mathbf{N}(I\downarrow i), \mathcal{M}\big)$ (see Proposition 26.1 (2)). We are going to show that this functor satisfies the assumptions of Proposition 23.5. The category $I\downarrow i$ clearly has a terminal object $t = (i, i \overset{id}{\to} i)$. Let us consider the following simplex $\sigma = (i_n \to \cdots \to i_1 \to i, i \overset{id}{\to} i)$ and morphism $\alpha = d_n \circ \cdots \circ d_1 : t \to \sigma$ in $\mathbf{N}(I\downarrow i)$. We have to show that α is sent via $colim_{\mathbf{N}(H\downarrow x)(-)}F$ to a weak equivalence. Since F is weakly equivalent to $\epsilon^* rF'$, the natural transformation $F(t,-) \to F(\sigma,-)$, induced by α, is a weak equivalence. These diagrams are cofibrant (see Proposition 26.1 (1)) and thus, on the colimits, they induce a weak equivalence $colim_{\mathbf{N}(H\downarrow x)(t)}F(t,-) \xrightarrow{\sim} colim_{\mathbf{N}(H\downarrow x)(\sigma)}F(\sigma,-)$. Proposition 23.5 therefore asserts that the following morphism is a weak equivalence as well:

$$colim_{\mathbf{N}(H(i)\downarrow x)}F(i,-) \xrightarrow{\sim} colim_{\mathbf{N}(I\downarrow i)}colim_{\mathbf{N}(H\downarrow x)(-)}F = \epsilon^k F(i,x)$$

The category $H(i)\downarrow x$ has a terminal object, the diagram $F(i,-)$ is bounded and cofibrant, and there is a weak equivalence $F(i,-) \to \epsilon^* rF'(i,-)$. Thus using Lemma 15.2, we get that:

$$colim_{\mathbf{N}(H(i)\downarrow x)}F(i,-) \to colim_{\mathbf{N}(H(i)\downarrow x)}\epsilon^* rF'(i,-) = rF'(i,x)$$

is a weak equivalence in $\mathcal{M}$. Combining the above two weak equivalences we can conclude that $\epsilon^k F(i,x) \to rF'(i,x)$ is also a weak equivalence.

As $\epsilon^k F(i,x)$ is a cofibrant object (see Proposition 26.1 (3)), the adjoint morphism $l\epsilon^k F(i,x) \to F'(i,x)$ is a weak equivalence in $\mathcal{C}$. We have thus checked that the indicated pair of adjoint functors forms a left model approximation.

Let us assume that $\mathcal{C}$ is closed under colimits. To show the second part of the theorem we need to prove that the following composite:

$$Fun^b\big(Gr_{\mathbf{N}(I)}\mathbf{N}(H), \mathcal{M}\big) \xrightarrow{\epsilon^k} Fun(Gr_I H, \mathcal{M}) \xrightarrow{l} Fun(Gr_I H, \mathcal{C}) \xrightarrow{colim_{Gr_I H}} \mathcal{C}$$

is homotopy meaningful on cofibrant objects. As left adjoints commute with colimits, and left Kan extensions do not modify them (see Proposition 36.2 (2)), this composite coincides with:

$$Fun^b\big(Gr_{\mathbf{N}(I)}\mathbf{N}(H),\mathcal{M}\big)\xrightarrow{colim_{Gr_{\mathbf{N}(I)}\mathbf{N}(H)}}\mathcal{M}\xrightarrow{l}\mathcal{C}$$

The second part of the theorem is now a consequence of the following two facts: both of the functors $colim_{Gr_{\mathbf{N}(I)}\mathbf{N}(H)}$ and l are homotopy meaningful on cofibrant objects (see Proposition 26.2) and $colim_{Gr_{\mathbf{N}(I)}\mathbf{N}(H)}$ preserves cofibrancy (see Proposition 26.1 (3)). □

26.7. COROLLARY. *Let $\mathcal{C}$ be closed under colimits and $l:\mathcal{M}\rightleftarrows\mathcal{C}:r$ be a left model approximation. Then the composite:*

$$\begin{array}{ccccc} Fun(Gr_IH,\mathcal{C}) & \xrightarrow{r} & Fun(Gr_IH,\mathcal{M}) & \xrightarrow{\epsilon^*} & Fun^b\big(Gr_{\mathbf{N}(I)}\mathbf{N}(H),\mathcal{M}\big) \\ & & & & \downarrow{\scriptstyle ocolim_{Gr_{\mathbf{N}(I)}\mathbf{N}(H)}} \\ & & Ho(\mathcal{C}) & \xleftarrow{l} & Ho(\mathcal{M}) \end{array}$$

is the total left derived functor of $colim_{Gr_IH}$. □

As a corollary of Proposition 26.5 and Corollary 26.7 we get the so-called Thomason Theorem for homotopy colimits:

26.8. THEOREM. *Let $H:I\to Cat$ be a diagram. Let $\mathcal{C}$ be a category closed under colimits and $l:\mathcal{M}\rightleftarrows\mathcal{C}:r$ be a left model approximation such that $\mathcal{M}$ has a functorial factorization of morphisms into cofibrations followed by acyclic fibrations. Then, for any $F:Gr_IH\to\mathcal{C}$, we have a weak equivalence in $\mathcal{C}$:*

$$hocolim_{Gr_IH}F\simeq hocolim_{i\in I}hocolim_{H(i)}F$$ □

26.9. COROLLARY (R. W. Thomason [**47**, Theorem 1.2]). *Let $H:I\to Cat$ be a diagram of small categories. Then there is a weak equivalence of spaces:*

$$N(Gr_IH)\simeq hocolim_IN(H)$$

PROOF. Apply Theorem 26.8 to the constant diagram $\Delta[0]:Gr_IH\to Spaces$. □

27. Étale spaces

In this section we discuss homotopy colimits of contravariant functors indexed by simplex categories, i.e., we look at functors of the form $F:\mathbf{K}^{op}\to\mathcal{C}$. We take advantage of the fact that after applying the nerve functor to a map of spaces $f:L\to K$ we get a reduced map $N(f^{op}):N(\mathbf{L}^{op})\to N(\mathbf{K}^{op})$ (a map which sends non-degenerate simplices in $N(\mathbf{L}^{op})$ to non-degenerate simplices in $N(\mathbf{K}^{op})$, see Example 12.10). The key property of such maps is that the pull-back process along them preserves absolute cofibrancy of bounded diagrams (see Proposition 12.12). The theory elaborated in this section will be used in Section 29 to define another way of computing homotopy colimits. This is then one of the ingredients in the proof of our cofinality result, Theorem 30.5.

Let $\mathcal{C}$ be a category closed under colimits and $l:\mathcal{M}\rightleftarrows\mathcal{C}:r$ be a left model approximation. Let K be a space. Recall that $\epsilon:\mathbf{N}(\mathbf{K}^{op})\to\mathbf{K}^{op}$ denotes the forgetful functor (see Definition 6.6). Throughout this section let us fix a functor $F:\mathbf{K}^{op}\to\mathcal{C}$ and a cofibrant replacement $QF\overset{\sim}{\twoheadrightarrow}\epsilon^*rF$ in $Fun^b\big(\mathbf{N}(\mathbf{K}^{op}),\mathcal{M}\big)$ of the composite $\mathbf{N}(\mathbf{K}^{op}) \xrightarrow{\epsilon} \mathbf{K}^{op} \xrightarrow{F} \mathcal{C} \xrightarrow{r} \mathcal{M}$. We are going to think about F as a fixed "system of coefficients" and study spaces over K by twisting them with F.

27.1. DEFINITION. Let $f : L \to K$ be a map. The colimit $colim_{\mathbf{L}^{op}}(f^{op})^*F$ is denoted simply by $colim_{\mathbf{L}^{op}}F$. The colimit $colim_{\mathbf{N}(\mathbf{L}^{op})}l\big(N(f^{op})^*QF\big)$ is denoted by $hocolim_{\mathbf{L}^{op}}F$ and called the *étale space* of the pull-back $(f^{op})^*F$.

In this way we get two functors:

$$colim\,\_\,F : Spaces{\downarrow}K \to \mathcal{C}\ ,\ \ (f : L \to K) \mapsto colim_{\mathbf{L}^{op}}F$$

$$hocolim\,\_\,F : Spaces{\downarrow}K \to \mathcal{C}\ ,\ \ (f : L \to K) \mapsto hocolim_{\mathbf{L}^{op}}F$$

The morphisms:

$$colim_{\mathbf{N}(\mathbf{L}^{op})}l\big(N(f^{op})^*QF\big) \to colim_{\mathbf{N}(\mathbf{L}^{op})}l\big((N(f^{op})^*\epsilon^*rF\big) \to colim_{\mathbf{L}^{op}}(f^{op})^*F$$

which are induced by $QF \overset{\sim}{\twoheadrightarrow} \epsilon^*rF$ and the adjointness of l and r form a natural transformation $hocolim\,\_\,F \to colim\,\_\,F$.

The notation $hocolim_{\mathbf{L}^{op}}F$, for the étale space of $(f^{op})^*F$, is justified by:

27.2. PROPOSITION. *Let $f : L \to K$ be a map. The étale space $hocolim_{\mathbf{L}^{op}}F$ is weakly equivalent to the homotopy colimit of the diagram $(f^{op})^*F : \mathbf{L}^{op} \to \mathcal{C}$.*

PROOF. Since the map $N(f^{op}) : N(\mathbf{L}^{op}) \to N(\mathbf{K}^{op})$ is reduced, the composite:

$$\mathbf{N}(\mathbf{L}^{op}) \xrightarrow{N(f^{op})} \mathbf{N}(\mathbf{K}^{op}) \xrightarrow{QF} \mathcal{M}$$

is a *cofibrant* replacement of:

$$\mathbf{N}(\mathbf{L}^{op}) \xrightarrow{\epsilon} \mathbf{L}^{op} \xrightarrow{f^{op}} \mathbf{K}^{op} \xrightarrow{F} \mathcal{C} \xrightarrow{r} \mathcal{M}$$

The proposition now follows from Corollary 16.2. □

An important property of the étale space construction is its additivity with respect to the indexing space. This property distinguishes the ocolimit construction from the hocolimit one (see Remark 14.3).

27.3. PROPOSITION. *Let $H : I \to Spaces{\downarrow}K$ be a functor. Then the natural morphism $colim_I hocolim_{\mathbf{H}^{op}}F \to hocolim_{(colim_I\mathbf{H})^{op}}F$ is an isomorphism in $\mathcal{C}$.*

PROOF. The proposition is a consequence of Proposition 8.2 (2) and the fact that $N\big((colim_I\mathbf{H})^{op}\big)$ is naturally isomorphic to $colim_I N(\mathbf{H}^{op})$ (see 6.11). □

When the considered left model approximation is given by the identity functors $id : \mathcal{M} \rightleftarrows \mathcal{M} : id$, the étale space construction converts cofibrations in *Spaces* into cofibrations in $\mathcal{M}$.

27.4. PROPOSITION. *Let $\mathcal{M}$ be a model category and $F : \mathbf{K}^{op} \to \mathcal{M}$ be a functor.*

1. *For any map $f : L \to K$, $hocolim_{\mathbf{L}^{op}}F$ is a cofibrant object in $\mathcal{M}$.*
2. *Let $h : N \to L$ be a map in $Spaces{\downarrow}K$. If h is a monomorphism, then $hocolim_{\mathbf{N}^{op}}F \to hocolim_{\mathbf{L}^{op}}F$ is a cofibration in $\mathcal{M}$.*

PROOF. Part 1 follows from Corollary 20.6 (2). Part 2 is a consequence of Corollary 20.2. □

We can combine Propositions 27.3 and 27.4 and get:

27.5. COROLLARY. *Let $\mathcal{M}$ be a model category and $F : \mathbf{K}^{op} \to \mathcal{M}$ be a functor.*

1. *Let the following be a push-out square of spaces over K, with the indicated arrows being cofibrations:*

$$\begin{array}{ccccc} A & \hookrightarrow & B & & \\ \downarrow & & \downarrow & & \\ C & \hookrightarrow & D & \longrightarrow & K \end{array}$$

Then the following is a push-out square in $\mathcal{M}$, with the indicated arrows being cofibrations:

$$\begin{array}{ccc} hocolim_{\mathbf{A}^{op}} F & \hookrightarrow & hocolim_{\mathbf{B}^{op}} F \\ \downarrow & & \downarrow \\ hocolim_{\mathbf{C}^{op}} F & \hookrightarrow & hocolim_{\mathbf{D}^{op}} F \end{array}$$

2. *Let the following be a telescope diagram of spaces over K:*

$$A \;=\; colim \;\Big(\begin{array}{ccccccc} A_0 & \hookrightarrow & A_1 & \hookrightarrow & A_2 & \hookrightarrow & \cdots \\ & \searrow & \downarrow & \swarrow & & & \\ & & K & & & & \end{array} \Big)$$

where, for $i \geq 0$, $A_i \hookrightarrow A_{i+1}$ is a cofibration. Then $hocolim_{\mathbf{A}^{op}} F$ is isomorphic to the telescope $colim\big(hocolim_{\mathbf{A_0}^{op}} F \hookrightarrow hocolim_{\mathbf{A_1}^{op}} F \hookrightarrow \cdots\big)$, where the indicated arrows are cofibrations. □

27.6. DEFINITION. Let $\mathcal{C}$ be a category with weak equivalences. We say that a functor $F : I \to \mathcal{C}$ is *homotopically constant* if any morphism in I is sent, via F, to a weak equivalence in $\mathcal{C}$.

We are going to apply Corollary 27.5 to calculate the étale space of a homotopically constant functor indexed by a contractible space.

27.7. PROPOSITION. *Let $F : \mathbf{K}^{op} \to \mathcal{C}$ be a homotopically constant functor. If $f : L \to K$ is a map such that L is contractible, then, for any simplex $\sigma \in L$, we have a weak equivalence $F\big(f(\sigma)\big) \simeq hocolim_{\mathbf{L}^{op}} F$ in $\mathcal{C}$.*

We first prove the proposition in the case when the considered model approximation is given by the identity functors $id : \mathcal{M} \rightleftarrows \mathcal{M} : id$.

27.8. LEMMA. *Let $\mathcal{M}$ be a model category and $F : \mathbf{K}^{op} \to \mathcal{M}$ be a homotopically constant functor. If $f : L \to K$ is a map such that L is contractible, then for any simplex $\sigma \in L$, the morphism:*

$$F\big(f(\sigma)\big) \simeq QF\big(f(\sigma)\big) \to colim_{\mathbf{N}(\mathbf{L}^{op})} N(f^{op})^* QF = hocolim_{\mathbf{L}^{op}} F$$

is a weak equivalence in $\mathcal{M}$.

PROOF. As F is homotopically constant, it is enough to show that for *some* simplex $\sigma \in L$, the morphism $QF\big(f(\sigma)\big) \to hocolim_{\mathbf{L}^{op}} F$ is a weak equivalence.

The proof is divided into several steps. In each step we show that the lemma is true for more and more complicated spaces L.

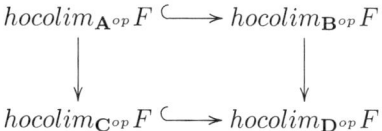

Step 1. Let $L = \Delta[n]$. The category $\mathbf{\Delta}[n]^{op}$ has an initial object ι (see Example 6.2). As F is homotopically constant, then so is $QF : \mathbf{N}(\mathbf{K}^{op}) \to \mathcal{M}$. It follows that the composite:

$$\mathbf{N}(\mathbf{\Delta}[n]^{op}) \xrightarrow{N(f^{op})} \mathbf{N}(\mathbf{K}^{op}) \xrightarrow{QF} \mathcal{M}$$

satisfies the assumptions of Proposition 23.7. We can therefore conclude that the following morphism is a weak equivalence in $\mathcal{M}$:

$$QF\big(f(\iota)\big) \to colim_{\mathbf{N}(\mathbf{\Delta}[n]^{op})} N(f^{op})^* QF = hocolim_{\mathbf{\Delta}[n]^{op}} F$$

Step 2. We say that a space A is basic if it can be built inductively starting with $\Delta[0]$ and gluing simplices along horns. Explicitly, one can present A as a telescope $colim(A_0 \hookrightarrow A_1 \hookrightarrow \cdots)$ with $A_0 = \Delta[0]$ and A_{n+1} obtained from A_n by a push-out $colim(A_n \leftarrow \coprod \Delta[n+1,k] \hookrightarrow \coprod \Delta[n+1])$. A basic space is necessarily contractible.

Let us assume that L is basic and finite dimensional. We are going to prove the lemma by induction on the dimension of L.

In the case $dim(L) = 0$, we have $L = \Delta[0]$ and thus Step 2 follows from Step 1. Let us assume that the proposition holds for those basic spaces whose dimension is less than n. Let L be basic and $dim(L) = n$. For simplicity assume that L has only one non-degenerate simplex of dimension n. In this case L fits into a push-out square:

$$\begin{array}{ccccc} \Delta[n,k] & \hookrightarrow & \Delta[n] & & \\ \downarrow & & \downarrow & & \\ N & \hookrightarrow & L & \longrightarrow & K \end{array}$$

where N is basic and $dim(N) < n$.

By Corollary 27.5 (1) we get a push-out diagram of étale spaces in $\mathcal{M}$:

$$\begin{array}{ccc} hocolim_{\mathbf{\Delta}[n,k]^{op}} F & \hookrightarrow & hocolim_{\mathbf{\Delta}[n]^{op}} F \\ \downarrow & & \downarrow \\ hocolim_{\mathbf{N}^{op}} F & \hookrightarrow & hocolim_{\mathbf{L}^{op}} F \end{array}$$

where the horizontal morphisms are cofibrations as indicated.

Since $\Delta[n,k]$ is basic and $dim(\Delta[n,k]) < n$, the inductive assumption implies that $QF\big(f(\sigma)\big) \to hocolim_{\mathbf{\Delta}[n,k]^{op}} F$ is a weak equivalence for any $\sigma \in \Delta[n,k]$. Therefore, according to Step 1, $hocolim_{\mathbf{\Delta}[n,k]^{op}} F \to hocolim_{\mathbf{\Delta}[n]^{op}} F$ is an *acyclic* cofibration. It follows that so is $hocolim_{\mathbf{N}^{op}} F \to hocolim_{\mathbf{L}^{op}} F$ (see Proposition 2.1). As by inductive assumption $QF\big(f(\sigma)\big) \to hocolim_{\mathbf{N}^{op}} F$ is a weak equivalence, we can conclude that $QF\big(f(\sigma)\big) \to hocolim_{\mathbf{L}^{op}} F$ is a weak equivalence as well.

Step 3. Let L be basic but not necessarily finite dimensional. In this case L can be presented as a colimit $L = colim(L_0 \hookrightarrow L_1 \hookrightarrow \cdots)$, where L_n is basic and n-dimensional. Step 3 follows now from Step 2 and Corollary 27.5 (2).

Step 4. Let L be contractible. This implies that $f : L \to K$ is a retract in $Spaces\!\downarrow\! K$ of a map $PL \to K$, where PL is basic. Explicitly, there are maps $L \to PL$ and $PL \to L$ in $Spaces\!\downarrow\! K$ for which the composite $L \to PL \to L$ is the identity.

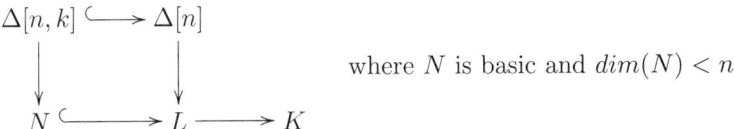

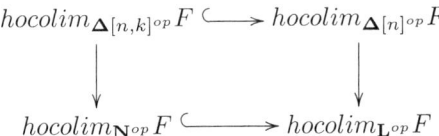

Let us choose a simplex $\sigma \in L$ and form the following commutative diagram:

$$\begin{array}{ccccc} QF\big(f(\sigma)\big) & \xrightarrow{id} & QF\big(f(\sigma)\big) & \xrightarrow{id} & QF\big(f(\sigma)\big) \\ \downarrow & & \downarrow & & \downarrow \\ hocolim_{\mathbf{L}^{op}}F & \longrightarrow & hocolim_{\mathbf{PL}^{op}}F & \longrightarrow & hocolim_{\mathbf{L}^{op}}F \end{array}$$

(the bottom composite $hocolim_{\mathbf{L}^{op}}F \to hocolim_{\mathbf{L}^{op}}F$ is id)

According to Step 3 the morphism $QF\big(f(\sigma)\big) \to hocolim_{\mathbf{PL}^{op}}F$ is a weak equivalence. It follows from the axiom **MC3** (see Section 2), that so is the morphism $QF\big(f(\sigma)\big) \to hocolim_{\mathbf{L}^{op}}F$. □

PROOF OF PROPOSITION 27.7. The proposition is a consequence of the lemma and Corollary 16.4. □

28. Diagrams indexed by cones II

Throughout Sections 28–30 we are going to fix a category $\mathcal{C}$, which is closed under colimits, and a left model approximation $l : \mathcal{M} \rightleftarrows \mathcal{C} : r$, such that $\mathcal{M}$ has a functorial factorization of morphisms into cofibrations followed by acyclic fibrations.

In this section we consider contravariant functors indexed by cones. We prove statements analogous to those in Section 22.

Let K be a space and CK be its cone (see Definition 22.1). The opposite category $(\mathbf{CK})^{op}$ can be represented as a Grothendieck construction (see Remark 22.3):

$$(\mathbf{CK})^{op} = Gr\big(\mathbf{K}^{op} \xleftarrow{p_1} \mathbf{K}^{op} \times \mathbf{\Delta}[0]^{op} \xrightarrow{p_2} \mathbf{\Delta}[0]^{op}\big)$$

where:

- $\mathbf{K}^{op}$ corresponds to the full subcategory of $(\mathbf{CK})^{op}$ consisting of simplices of the form (σ, e^0).
- $\mathbf{K}^{op} \times \mathbf{\Delta}[0]^{op}$ corresponds to the full subcategory of $(\mathbf{CK})^{op}$ consisting of simplices of the form (σ, e^l), where $l \geq 1$.
- $\mathbf{\Delta}[0]^{op}$ corresponds to the full subcategory of $(\mathbf{CK})^{op}$ consisting of simplices of the form e^l.

28.1. PROPOSITION. *Let $F : (\mathbf{CK})^{op} \to \mathcal{C}$ be a functor. Assume that F sends morphisms of the form $d_j : (\sigma, e^l) \to (\sigma, e^{l-1})$ (for $j > dim(\sigma)$) and $d_j : e^l \to e^{l-1}$ (for $j \geq 0$) to weak equivalences in $\mathcal{C}$. Then $hocolim_{(\mathbf{CK})^{op}}F$ is weakly equivalent to $F(e^1)$.*

PROOF. To prove the proposition we use Thomason's theorem (see Theorem 26.8). It gives a weak equivalence:

$$hocolim_{(\mathbf{CK})^{op}}F \;\simeq\; hocolim \left(\begin{array}{ccc} hocolim_{\mathbf{K}^{op}\times\mathbf{\Delta}[0]^{op}}F & \longrightarrow & hocolim_{\mathbf{\Delta}[0]^{op}}F \\ \downarrow & & \\ hocolim_{\mathbf{K}^{op}}F & & \end{array} \right)$$

According to the Fubini theorem (see Theorem 24.9):

$$hocolim_{\mathbf{K}^{op}\times\mathbf{\Delta}[0]^{op}}F \simeq hocolim_{\mathbf{K}^{op}} hocolim_{\mathbf{\Delta}[0]^{op}}F$$

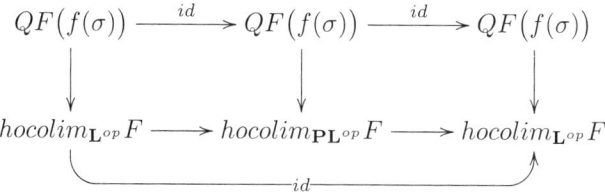

Let σ be an object in $\mathbf{K}^{op}$. We denote by $F(\sigma,-):\mathbf{\Delta}[0]^{op}\to\mathcal{C}$ the following composite:

$$\{\sigma\}\times\mathbf{\Delta}[0]^{op}\subset\mathbf{K}^{op}\times\mathbf{\Delta}[0]^{op}\subset(\mathbf{CK})^{op}\xrightarrow{F}\mathcal{C}$$

The assumptions on F imply that $F(\sigma,-):\mathbf{\Delta}[0]^{op}\to\mathcal{C}$ is homotopically constant. Thus, by Proposition 27.7, $hocolim_{\mathbf{\Delta}[0]^{op}}F(\sigma,-)\simeq F(\sigma,e^0)$. We can conclude that the functor $\mathbf{K}^{op}\ni\sigma\mapsto hocolim_{\mathbf{\Delta}[0]^{op}}F(\sigma,-)$ is weakly equivalent to the composite $\mathbf{K}^{op}\subset(\mathbf{CK})^{op}\xrightarrow{F}\mathcal{C}$. The morphism:

$$hocolim_{\mathbf{K}^{op}}F(-,e^0)\to hocolim_{\mathbf{K}^{op}}hocolim_{\mathbf{\Delta}[0]^{op}}F$$

is therefore a weak equivalence, and hence so is:

$$hocolim_{\mathbf{K}^{op}\times\mathbf{\Delta}[0]^{op}}F\to hocolim_{\mathbf{K}^{op}}F$$

This argument shows that $hocolim_{(\mathbf{CK})^{op}}F\simeq hocolim_{\mathbf{\Delta}[0]^{op}}F$. However, since the functor $F:\mathbf{\Delta}[0]^{op}\to\mathcal{C}$ is also homotopically constant (it is a consequence of the assumptions on F), we get a weak equivalence $F(e^1)\simeq hocolim_{(\mathbf{CK})^{op}}F$. □

Proposition 28.1 can be applied to identify homotopy colimits of certain contravariant functors indexed by nerves of categories with terminal objects. Recall that $\epsilon:\mathbf{N}(CI)^{op}\to CI$, $(i_n\to\cdots\to i_0)\mapsto i_n$ denotes the forgetful functor (see Definition 6.6).

28.2. Corollary. *Let $F:\mathbf{N}(CI)^{op}\to\mathcal{C}$ be a functor. Assume that there exists $F':CI\to\mathcal{C}$ and a weak equivalence $F\xrightarrow{\sim}\epsilon^*F'$ in $Fun(\mathbf{N}(CI)^{op},\mathcal{C})$. Then the composite:*

$$\begin{array}{ccc} hocolim_{\mathbf{N}(CI)^{op}}F \longrightarrow & hocolim_{\mathbf{N}(CI)^{op}}\epsilon^*F' & \\ & \downarrow & \\ & colim_{\mathbf{N}(CI)^{op}}\epsilon^*F' & \longrightarrow colim_{CI}F'=F'(e) \end{array}$$

is a weak equivalence in $\mathcal{C}$. In particular $hocolim_{\mathbf{N}(CI)^{op}}F\simeq F(e)$. □

Corollary 28.2 can be generalized to an arbitrary small category with a terminal object.

28.3. Proposition. *Let I be a small category with a terminal object t and $F:\mathbf{N}(I)^{op}\to\mathcal{C}$ be a functor. Assume that there exists $F':I\to\mathcal{C}$ and a weak equivalence $F\xrightarrow{\sim}\epsilon^*F'$ in $Fun(\mathbf{N}(I)^{op},\mathcal{C})$. Then the composite:*

$$hocolim_{\mathbf{N}(I)^{op}}F\to hocolim_{\mathbf{N}(I)^{op}}\epsilon^*F'\to colim_{\mathbf{N}(I)^{op}}\epsilon^*F'\to colim_IF'=F'(t)$$

is a weak equivalence in $\mathcal{C}$. In particular $hocolim_{\mathbf{N}(I)^{op}}F\simeq F(t)$.

Proof. The proposition can be proven using Corollary 28.2 and the fact that in the case I has a terminal object, it is a retract of its cone CI (compare with the proof of Proposition 23.5). □

29. Homotopy colimits as étale spaces

In this section we show that the forgetful functor $\epsilon : \mathbf{N}(I)^{op} \to I$ is cofinal with respect to taking homotopy colimits. This reduces calculating homotopy colimits indexed by any small category to calculating étale spaces.

29.1. THEOREM. *The composite:*

$$Fun(I, \mathcal{C}) \xrightarrow{\epsilon^*} Fun(\mathbf{N}(I)^{op}, \mathcal{C}) \xrightarrow{hocolim_{\mathbf{N}(I)^{op}}} Ho(\mathcal{C})$$

together with the natural transformation induced by the morphisms:

$$hocolim_{\mathbf{N}(I)^{op}}\epsilon^* F \to colim_{\mathbf{N}(I)^{op}}\epsilon^* F = colim_I F$$

is the total left derived functor of $colim_I$.

PROOF. It is clear that this composite is homotopy invariant. Thus to prove the theorem we have to show that, if $\mathcal{G} : Fun(I, \mathcal{C}) \to Ho(\mathcal{C})$ is also a homotopy invariant functor, then any natural transformation $\mathcal{G} \to colim_I$ factors uniquely as $\mathcal{G} \to hocolim_{\mathbf{N}(I)^{op}}\epsilon^*(-) \to colim_I$. For any $F \in Fun(I, \mathcal{C})$ we need to define a morphism $\mathcal{G}(F) \to hocolim_{\mathbf{N}(I)^{op}}\epsilon^* F$ in $Ho(\mathcal{C})$.

Let $F : I \to \mathcal{C}$ be a functor. The nerve $N(I)$ can be expressed as a colimit $colim_I N(I \downarrow -)$ (see Example 6.7). As the étale space construction is additive with respect to the indexing space (see Proposition 27.3), we get an isomorphism in $\mathcal{C}$:

$$hocolim_{\mathbf{N}(I)^{op}}\epsilon^* F = colim_I hocolim_{\mathbf{N}(I \downarrow -)^{op}}\epsilon^* F$$

Consider the diagram $I \ni i \mapsto hocolim_{\mathbf{N}(I \downarrow i)^{op}}\epsilon^* F \in \mathcal{C}$. As the category $I \downarrow i$ has a terminal object, Proposition 28.3 asserts that we have a weak equivalence:

$$hocolim_{\mathbf{N}(I \downarrow i)^{op}}\epsilon^* F \to colim_{\mathbf{N}(I \downarrow i)^{op}}\epsilon^* F = colim_{I \downarrow i} F = F(i)$$

Let us denote by $\Psi : hocolim_{\mathbf{N}(I \downarrow -)^{op}}\epsilon^* F \to F$ the natural transformation induced by these morphisms. Since it is a weak equivalence and $\mathcal{G}$ is homotopy invariant, $\mathcal{G}(\Psi)$ is an isomorphism in $Ho(\mathcal{C})$. We can now define $\mathcal{G}(F) \to hocolim_{\mathbf{N}(I)^{op}}\epsilon^* F$ to be the following composite in $Ho(\mathcal{C})$:

$$\begin{array}{ccc} \mathcal{G}(F) \xrightarrow{\mathcal{G}(\Psi)^{-1}} \mathcal{G}(hocolim_{\mathbf{N}(I \downarrow -)^{op}}\epsilon^* F) & & \\ \downarrow & & \\ colim_I hocolim_{\mathbf{N}(I \downarrow -)^{op}}\epsilon^* F & = & hocolim_{\mathbf{N}(I)^{op}}\epsilon^* F \end{array}$$

In this way we get the desired natural transformation $\mathcal{G} \to hocolim_{\mathbf{N}(I)^{op}}\epsilon^*(-)$. Its uniqueness is clear. □

29.2. COROLLARY. *Let $F : I \to \mathcal{C}$ be a homotopically constant functor. If I is contractible, then $hocolim_I F$ is weakly equivalent to $F(i)$ for any $i \in I$.*

PROOF. The corollary is a consequence of Theorem 29.1 and Proposition 27.7. □

30. Cofinality

In this section we discuss cofinality properties of the homotopy colimit construction. We generalize [**7**, Theorem XI.9.2] to categories with left model approximations. A similar result is also proven in [**18**, 10.7].

We give a sufficient condition for a functor $f : J \to I$ to induce a weak equivalence $hocolim_J f^*F \to hocolim_I F$ for any diagram $F : I \to \mathcal{C}$.

30.1. DEFINITION. Let I and J be small categories. A functor $f : J \to I$ is said to be *terminal* if the space $N(i \downarrow f)$ is contractible for every $i \in I$.

Terminal functors are cofinal with respect to taking colimits, i.e., if $f : J \to I$ is terminal, then, for any diagram $F : I \to \mathcal{C}$, the morphism $colim_J f^*F \to colim_I F$ is an isomorphism (see Proposition 37.1).

30.2. EXAMPLE. Consider the functor $J \to *$. The category $* \downarrow J$ can be identified with J. Thus $J \to *$ is terminal if and only if J is contractible.

30.3. EXAMPLE. Let I be a category with a terminal object t. The functor $* \to I$, $* \mapsto t$, is terminal. Indeed, for any $i \in I$, the category $i \downarrow *$ is the trivial one.

30.4. EXAMPLE. The forgetful functor $\epsilon : \mathbf{N}(I)^{op} \to I$, $(i_n \to \cdots \to i_0) \mapsto i_n$, is terminal.

30.5. THEOREM. *Let $f : J \to I$ be terminal. Then the composite:*

$$Fun(I, \mathcal{C}) \xrightarrow{f^*} Fun(J, \mathcal{C}) \xrightarrow{hocolim_J} Ho(\mathcal{C})$$

together with the natural transformation induced by the morphisms:

$$hocolim_J f^*F \to colim_J f^*F = colim_I F$$

is the total left derived functor of $colim_I$. Explicitly, there is a natural weak equivalence $hocolim_J f^*F \simeq hocolim_I F$.

PROOF. We start with describing two functors:

- $H : \mathbf{N}(I)^{op} \to Cat$ is defined as $H(i_n \to \cdots \to i_0) := i_0 \downarrow f$.
- $G : J \to Cat$ is defined as $G(j) := \mathbf{N}\big(I \downarrow f(j)\big)^{op}$.

Consider their Grothendieck constructions $Gr_{\mathbf{N}(I)^{op}} H$ and $Gr_J G$. Observe that the following is an isomorphism of categories:

$$Gr_{\mathbf{N}(I)^{op}} H \to Gr_J G$$

$$\big((i_n \to \cdots \to i_0); (j, i_0 \xrightarrow{\alpha} f(j))\big) \longmapsto \big(j\,;\, i_n \to \cdots \to i_0 \xrightarrow{\alpha} f(j)\big)$$

We are going to use the symbol Λ to denote the category $Gr_{\mathbf{N}(I)^{op}} H = Gr_J G$ and λ to denote the composite $\Lambda = Gr_{\mathbf{N}(I)^{op}} H \to \mathbf{N}(I)^{op} \xrightarrow{\epsilon} I$, where $\epsilon : \mathbf{N}(I)^{op} \to I$ is the forgetful functor.

Let $F : I \to \mathcal{C}$ be a functor. We are going to apply Thomason's theorem (see Theorem 26.8) to calculate the homotopy colimit of $\lambda^* F : \Lambda \to \mathcal{C}$. We can do it using the two presentations of the category Λ as a Grothendieck construction.

Case $\Lambda = Gr_{\mathbf{N}(I)^{op}} H$. Theorem 26.8 asserts that there is a weak equivalence:

$$hocolim_\Lambda \lambda^* F \simeq hocolim_{\mathbf{N}(I)^{op}} hocolim_H \lambda^* F$$

For any $\sigma = (i_n \to \cdots \to i_0) \in N(I)^{op}$, the diagram $\lambda^*F : H(\sigma) = i_0 \downarrow f \to \mathcal{C}$ is constant with value $F(i_n)$. Since by assumption the category $i_0 \downarrow f$ is contractible, Proposition 27.7 implies that the following morphism is a weak equivalence in $\mathcal{C}$:

$$hocolim_{H(\sigma)}\lambda^*F \to colim_{i_0 \downarrow f} F(i_n) = F(i_n)$$

The object $F(i_n)$ can be identified with $\epsilon^*F(\sigma)$. Thus the induced natural transformation $hocolim_{H(-)}\lambda^*F \to \epsilon^*F$ is a weak equivalence, and hence so is the morphism:

$$hocolim_{\mathbf{N}(I)^{op}} hocolim_H \lambda^*F \to hocolim_{\mathbf{N}(I)^{op}} \epsilon^*F$$

However, since $hocolim_{\mathbf{N}(I)^{op}}\epsilon^*F \simeq hocolim_I F$ (see Theorem 29.1), we can conclude that $hocolim_\Lambda \lambda^*F$ is naturally weakly equivalent to $hocolim_I F$.

Case $\Lambda = Gr_J G$**.** Theorem 26.8 asserts that there is a weak equivalence:

$$hocolim_\Lambda \lambda^*F \simeq hocolim_J hocolim_G \lambda^*F$$

For any $j \in J$, the diagram $\lambda^*F : G(j) = \mathbf{N}\big(I \downarrow f(j)\big)^{op} \to \mathcal{C}$ coincides with the composite $\mathbf{N}\big(I \downarrow f(j)\big)^{op} \xrightarrow{\epsilon} I \downarrow f(j) \to I \xrightarrow{F} \mathcal{C}$. Since the category $I \downarrow f(j)$ has a terminal object $f(j) \xrightarrow{id} f(j)$, according to Proposition 28.3, the following morphism is a weak equivalence in $\mathcal{C}$:

$$hocolim_{G(j)}\lambda^*F \to colim_{\mathbf{N}(I \downarrow f(j))^{op}}\epsilon^*F = colim_{I \downarrow f(j)} F = F\big(f(j)\big) = f^*F(j)$$

Thus the induced natural transformation $hocolim_{G(-)}\lambda^*F \to f^*F$ is also a weak equivalence, and hence so is the morphism $hocolim_J hocolim_G \lambda^*F \to hocolim_J f^*F$.

The theorem has been proven since we have found natural weak equivalences

$$hocolim_I F \simeq hocolim_\Lambda \lambda^*F \simeq hocolim_J f^*F$$ □

31. Homotopy limits

All the material presented in the entire paper can be dualized. In this section we are going to present an overview of some of those dual notions.

Let $F : \mathcal{D} \to \mathcal{C}$ and $G : \mathcal{D} \to \mathcal{C}$ be functors and $\Psi : F \to G$ be a natural transformation. The induced functor and natural transformation on the opposite categories are denoted respectively by $F^\vee : \mathcal{D}^{op} \to \mathcal{C}^{op}$ and $\Psi^\vee : G^\vee \to F^\vee$, and called the duals of F and Ψ.

31.1. *Right derived functors.* Let $\mathcal{D}$ be a category with weak equivalences and $\mathcal{H} : \mathcal{D} \to \mathcal{E}$ be a functor. A functor $R(\mathcal{H}) : \mathcal{D} \to \mathcal{E}$ together with a natural transformation $\mathcal{H} \to R(\mathcal{H})$ is called the right derived functor of $\mathcal{H}$ if the induced functor on the opposite categories $R(\mathcal{H})^\vee : \mathcal{D}^{op} \to \mathcal{E}^{op}$ together with the natural transformation $R(\mathcal{H})^\vee \to \mathcal{H}^\vee$ is the left derived functor of $\mathcal{H}^\vee : \mathcal{D}^{op} \to \mathcal{E}^{op}$. Explicitly, $R(\mathcal{H})$ sends weak equivalences in $\mathcal{D}$ to isomorphisms in $\mathcal{E}$, and the natural transformation $\mathcal{H} \to R(\mathcal{H})$ is initial with respect to this property.

If $\mathcal{C}$ is a category with weak equivalence that admits a localization $\mathcal{C} \to Ho(\mathcal{C})$, then the right derived functor of the composite $\mathcal{D} \xrightarrow{\mathcal{H}} \mathcal{C} \to Ho(\mathcal{C})$ is called the total right derived functor of $\mathcal{H}$.

Let $\mathcal{M}$ be a model category and $\mathcal{C}$ be a category with weak equivalences. We say that a functor $\mathcal{H} : \mathcal{M} \to \mathcal{C}$ is homotopy meaningful on fibrant objects if for any weak equivalence $f : X \to Y$, between fibrant objects in $\mathcal{M}$, $\mathcal{H}(f)$ is a weak equivalence in $\mathcal{C}$. Assume that $\mathcal{C}$ admits a localization. Then for any functor $\mathcal{H} : \mathcal{M} \to \mathcal{C}$, which is homotopy meaningful on fibrant objects, the total right derived functor exists. It can be constructed by assigning to $X \in \mathcal{M}$ the object $\mathcal{H}(RX)$ in $Ho(\mathcal{C})$, where RX is a fibrant replacement of X in $\mathcal{M}$.

31.2. *Right model approximations.* Let $\mathcal{D}$ be a category with weak equivalences. A model category $\mathcal{M}$ together with a pair of adjoint functors $l : \mathcal{D} \rightleftarrows \mathcal{M} : r$ is called a right model approximation if the induced functors $r^{\vee} : \mathcal{M}^{op} \rightleftarrows \mathcal{D}^{op} : l^{\vee}$ form a left model approximation as defined in Definition 5.1. Explicitly, the following conditions have to be satisfied:

1. the functor r is right adjoint to l;
2. the functor l is homotopy meaningful, i.e., if f is a weak equivalence in $\mathcal{D}$, then lf is a weak equivalence in $\mathcal{M}$;
3. the functor r is homotopy meaningful on fibrant objects;
4. for any object A in $\mathcal{D}$ and any fibrant object X in $\mathcal{M}$, if a morphism $lA \to X$ is a weak equivalence in $\mathcal{M}$, then so is its adjoint $A \to rX$ in $\mathcal{D}$.

As in the case of left approximations (see Proposition 5.5), if $\mathcal{D}$ has a right model approximation, then the localization of $\mathcal{D}$ with respect to weak equivalences exists.

Let $\mathcal{C}$ be a category with weak equivalences that admits a localization. We say that a right model approximation $l : \mathcal{D} \rightleftarrows \mathcal{M} : r$ is good for a functor $\mathcal{H} : \mathcal{D} \to \mathcal{C}$, if the composite $\mathcal{H} \circ r : \mathcal{M} \to \mathcal{C}$ is homotopy meaningful on fibrant objects. In such a case the total left derived functor of $\mathcal{H}$ exists and can be constructed by assigning to $X \in \mathcal{D}$ the object $\mathcal{H}(rRlX) \in Ho(\mathcal{C})$, where RlX is a fibrant replacement of lX in $\mathcal{M}$.

31.3. *Bounded diagrams.* A functor $F : \mathbf{K}^{op} \to \mathcal{C}$ is called bounded if its dual $F^{\vee} : \mathbf{K} \to \mathcal{C}^{op}$ is bounded in the sense of Definition 17.1. The full subcategory of $Fun(\mathbf{K}^{op}, \mathcal{C})$ consisting of bounded diagrams is denoted by $Fun^b(\mathbf{K}^{op}, \mathcal{C})$.

Let $f : L \to K$ be a map and $\mathcal{C}$ be closed under limits. By definition, the right Kan extension $f_k : Fun(\mathbf{L}^{op}, \mathcal{C}) \to Fun(\mathbf{K}^{op}, \mathcal{C})$ is the dual of the left Kan extension $f^k : Fun(\mathbf{L}, \mathcal{C}^{op}) \to Fun(\mathbf{K}, \mathcal{C}^{op})$, i.e., $f_k = (f^k)^{\vee}$. It turns out that the functor f_k converts bounded diagrams into bounded diagrams and hence it induces a functor $f_k : Fun^b(\mathbf{L}^{op}, \mathcal{C}) \to Fun^b(\mathbf{K}^{op}, \mathcal{C})$, compare with Theorem 33.1. It is the right adjoint to the pull-back process $f^* : Fun^b(\mathbf{K}^{op}, \mathcal{C}) \to Fun^b(\mathbf{L}^{op}, \mathcal{C})$, $F \mapsto F \circ f$.

Let $\mathcal{M}$ be a model category. The category $Fun^b(\mathbf{K}^{op}, \mathcal{M})$ can be given a model structure where weak equivalences are the objectwise weak equivalences and cofibrations are the objectwise cofibrations. A natural transformation $\Psi : F \to G$ in $Fun^b(\mathbf{K}^{op}, \mathcal{M})$ is a fibration if $\Psi^{\vee} : G^{\vee} \to F^{\vee}$ is a cofibration in $Fun^b(\mathbf{K}, \mathcal{M}^{op})$. Explicitly, a bounded diagram $F : \mathbf{K}^{op} \to \mathcal{M}$ is fibrant if and only if, for any non-degenerate simplex $\sigma : \Delta[n] \to K$, the morphism $F(\sigma) \to lim_{\partial\mathbf{\Delta}[n]^{op}} F$ is a fibration in $\mathcal{M}$.

The limit functor $lim_{\mathbf{K}^{op}} : Fun^b(\mathbf{K}^{op}, \mathcal{M}) \to \mathcal{M}$ and the right Kan extension $f_k : Fun^b(\mathbf{L}^{op}, \mathcal{M}) \to Fun^b(\mathbf{K}^{op}, \mathcal{M})$ along $f : L \to K$ are homotopy meaningful on fibrant objects. Moreover they convert (acyclic) fibrations into (acyclic) fibrations, compare with Proposition 13.3. In particular if $F : \mathbf{K}^{op} \to \mathcal{M}$ is fibrant in $Fun^b(\mathbf{K}^{op}, \mathcal{M})$, then $lim_{\mathbf{K}^{op}} F$ is fibrant in $\mathcal{M}$.

31.4. *Bousfield-Kan approximation.* Let $l : \mathcal{C} \rightleftarrows \mathcal{M} : r$ be a right model approximation and I be a small category. Recall that $\epsilon : \mathbf{N}(I)^{op} \to I$ denotes the forgetful functor (see Definition 6.6). The pair of adjoint functors:

$$Fun(I,\mathcal{C}) \underset{r\circ\epsilon_k}{\overset{\epsilon^*\circ l}{\rightleftarrows}} Fun^b(\mathbf{N}(I)^{op},\mathcal{M})$$

is a right model approximation. It is called the Bousfield-Kan approximation of $Fun(I,\mathcal{C})$, as in the dual case Definition 11.1.

Let $f : I \to J$ be a functor. The Bousfield-Kan approximation is good for the functors $lim_I : Fun(I,\mathcal{C}) \to \mathcal{C}$ and $f_k : Fun(I,\mathcal{C}) \to Fun(J,\mathcal{C})$. In particular their right derived functors (the homotopy limit and the homotopy right Kan extension) exist. Let $F : I \to \mathcal{C}$, be a functor. Its homotopy limit $holim_I F$ can be identified with the homotopy colimit of its dual $hocolim_{I^{op}} F^\vee$.

31.5. *Fubini and Thomason theorems.* Let $H : I \to Cat$ be a functor. The co-Grothendieck construction $Gr^I H$ is by definition the following category:

- an object in $Gr^I H$ is a pair (i,a) consisting of an object $i \in I$ and an object $a \in H(i)$;
- a morphism $(\alpha,h) : (i,a) \to (j,b)$ in $Gr^I H$ is a pair (α,h) consisting of a morphism $\alpha : j \to i$ in I and a morphism $h : a \to H(\alpha)(b)$ in $H(i)$.
- the composite of $(\alpha,h) : (i,a) \to (j,b)$ and $(\beta,g) : (j,b) \to (l,c)$ is defined to be $(\alpha\circ\beta, H(\alpha)(g)\circ h)$.

If $J : I \to Cat$ is the constant functor with value J, then $Gr^I J$ can be identified with the product $I^{op} \times J$ in Cat.

Let $H : I \to Cat$ be a functor. The category $(Gr^I H)^{op}$ can be identified with the Grothendieck construction $Gr_I H^{op}$ of the composite of H and the "opposite category" functor $Cat \to Cat$, $J \mapsto J^{op}$.

To describe a functor $F : Gr^I H \to \mathcal{C}$ it is necessary and sufficient to have the following data:

1. a functor $F_i : H(i) \to \mathcal{C}$ for every object $i \in I$;
2. a natural transformation $F_\alpha : H(\alpha)^* F_i \to F_j$ for every morphism $\alpha : j \to i$ in I;
3. if $\alpha : j \to i$ and $\beta : i \to l$ are composable morphisms in I, then the natural transformations $F_{\beta\circ\alpha}$ and $F_\alpha \circ H(\alpha)^* F_\beta$ should coincide.

It follows that any $F : Gr^I H \to \mathcal{C}$ induces a new functor $lim_{H(-)} : I^{op} \to \mathcal{C}$, $i \mapsto lim_{H(i)} F$.

Let $\mathcal{C}$ be closed under limits and $l : \mathcal{C} \rightleftarrows \mathcal{M} : r$ be a right model approximation such that $\mathcal{M}$ has a functorial factorization of morphisms into acyclic cofibrations followed by fibrations. Then for any functor $F : Gr^I H \to \mathcal{C}$, the homotopy inverse limit $holim_{Gr^I H} F$ is weakly equivalent to $holim_{i\in I^{op}} holim_{H(i)} F$, compare with Theorem 26.8. In particular for any $F : I \times J \to \mathcal{C}$ there are weak equivalences in $\mathcal{C}$: $holim_{I\times J} F \simeq holim_I holim_J F \simeq holim_J holim_I F$.

31.6. *Cofinality.* A functor $f : J \to I$ is called initial if its dual $f^\vee : J^{op} \to I^{op}$ is terminal (see Definition 30.1). Explicitly, $f : J \to I$ is initial if for any $i \in I$, the space $N(f\downarrow i)$ is contractible. The forgetful functor $\epsilon : \mathbf{N}(I) \to I$ is an example of an initial functor.

Let $\mathcal{C}$ be closed under limits and $l : \mathcal{C} \rightleftarrows \mathcal{M} : r$ be a right model approximation such that $\mathcal{M}$ has a functorial factorization of morphisms into acyclic cofibrations followed by fibrations. If $f : J \to I$ is an initial functor, then for any functor

$F : I \to \mathcal{C}$, the morphism $holim_I F \to holim_J f^* F$ is a weak equivalence, compare with Theorem 30.5.

APPENDIX A

Left Kan extensions preserve boundedness

32. Degeneracy Map

In this section we investigate the degeneracy maps $s_i : \Delta[n+1] \to \Delta[n]$. More precisely, we discuss some properties of the fiber diagram $ds_i : \mathbf{\Delta}[n] \to Spaces$, associated to the map s_i (see Section 9). These will be used in the next section to study the behavior of bounded diagrams under the left Kan extension.

Observe that any map $\sigma : \Delta[m] \to \Delta[n]$ can be realized canonically as the nerve of a functor $[m] \to [n]$ (see Remark 6.4), which is also denoted by σ. For example, $s_i : \Delta[n+1] \to \Delta[n]$ is the nerve of the functor $s_i : [n+1] \to [n]$ determined by the assignment:

$$
\begin{array}{ccc}
[n+1] & \quad n+1 \twoheadrightarrow \cdots \twoheadrightarrow i+2 \twoheadrightarrow i+1 \\
\downarrow s_i & \qquad\qquad\qquad\qquad\qquad\qquad \downarrow \\
 & \qquad\qquad\qquad\qquad\qquad i \to i-1 \twoheadrightarrow \cdots \twoheadrightarrow 0 \\
[n] & \quad n \to \cdots \to i+1 \to i \to i-1 \twoheadrightarrow \cdots \twoheadrightarrow 0
\end{array}
$$

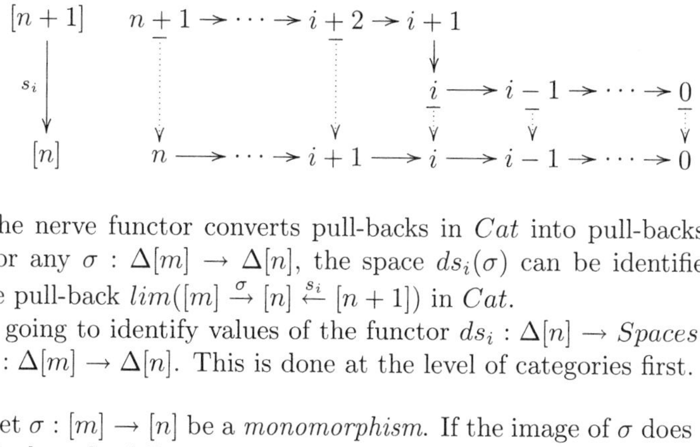

Since the nerve functor converts pull-backs in Cat into pull-backs in $Spaces$ (see 6.8), for any $\sigma : \Delta[m] \to \Delta[n]$, the space $ds_i(\sigma)$ can be identified with the nerve of the pull-back $lim([m] \xrightarrow{\sigma} [n] \xleftarrow{s_i} [n+1])$ in Cat.

We are going to identify values of the functor $ds_i : \mathbf{\Delta}[n] \to Spaces$ for various simplices $\sigma : \Delta[m] \to \Delta[n]$. This is done at the level of categories first.

32.1. Let $\sigma : [m] \to [n]$ be a *monomorphism*. If the image of σ does not contain the object i, then the following is a pull-back square in Cat:

$$
\begin{array}{ccc}
[m] & \longrightarrow & [n+1] \\
{\scriptstyle id}\downarrow & & \downarrow {\scriptstyle s_i} \\
[m] & \xrightarrow{\sigma} & [n]
\end{array}
$$

If the image of σ does contain the object i, then, for a certain j, we get a pull-back square in Cat of the form:

$$
\begin{array}{ccc}
[m+1] & \longrightarrow & [n+1] \\
{\scriptstyle s_j}\downarrow & & \downarrow {\scriptstyle s_i} \\
[m] & \xrightarrow{\sigma} & [n]
\end{array}
$$

32.2. Let us consider a degeneracy $s_j : [n+1] \to [n]$ where, for simplicity, we assume that $j < i$. Observe that the following is a pull-back square in Cat:

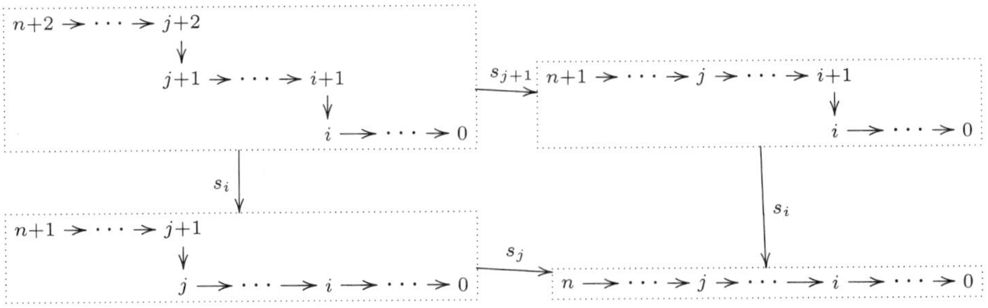

In short we have a pull-back square:

$$\begin{array}{ccc} [n+2] & \xrightarrow{s_{j+1}} & [n+1] \\ {\scriptstyle s_i}\downarrow & & \downarrow{\scriptstyle s_i} \\ [n+1] & \xrightarrow{s_j} & [n] \end{array}$$

32.3. Let us consider the simplex $\sigma : \Delta[m] \to \Delta[n]$ given by the composite $s_i \circ \cdots \circ s_{i+k-1} \circ s_{i+k} = (s_i)^{k+1}$. At the level of small categories this map corresponds to the functor given by the assignment:

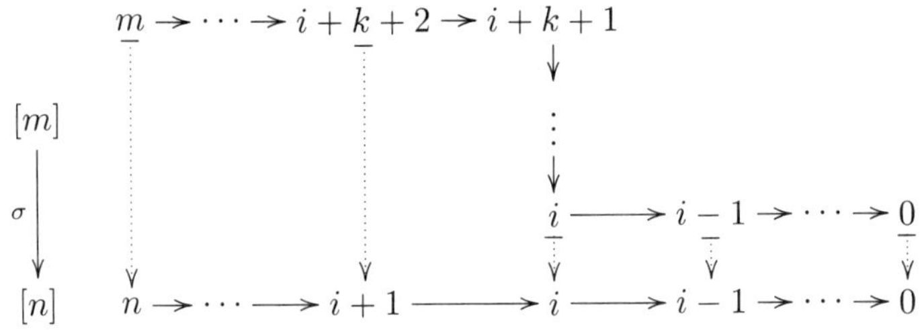

Observe that the following is a pull-back square in Cat:

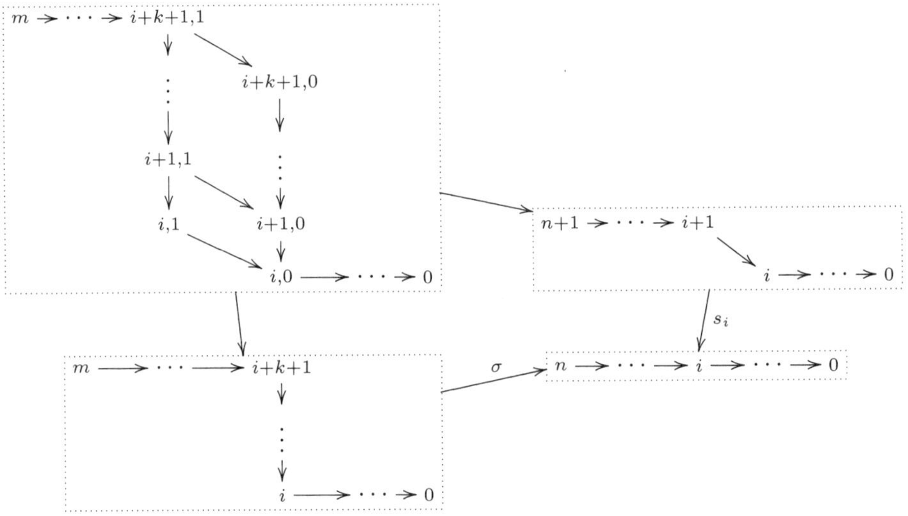

By P let us denote the category that sits in the top left corner of the above diagram. Consider subspaces of $N(P)$ which correspond to the subcategories of P given by the graphs:

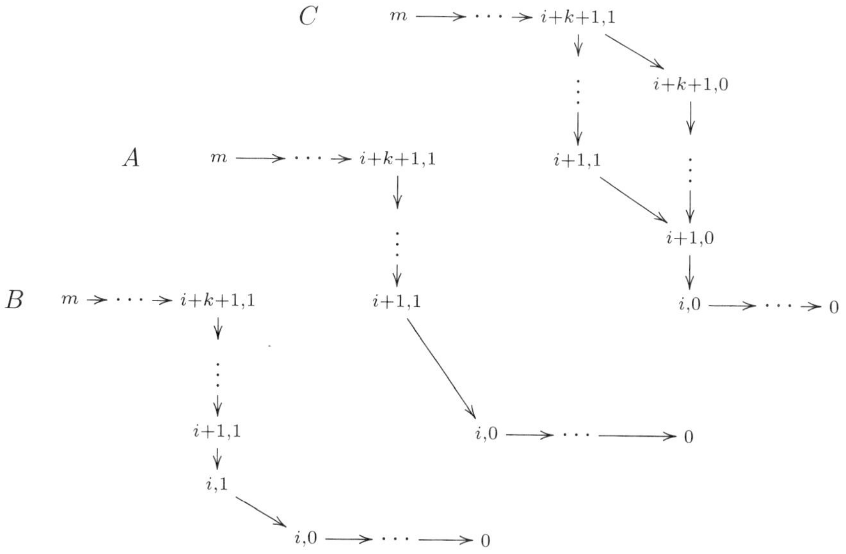

We can identify $N(B)$ with $\Delta[m+1]$, $N(A)$ with $\Delta[m]$, and the map $N(A) \to N(B)$ (induced by the inclusion $A \hookrightarrow B$) with $d_{i+1} : \Delta[m] \to \Delta[m+1]$. By induction we can identify $N(C)$ with the space that fits into the pull-back square:

$$\begin{array}{ccc} N(C) & \longrightarrow & \Delta[n+2] \\ \downarrow & & \downarrow{\scriptstyle s_{i+1}} \\ \Delta[m] & \xrightarrow{\sigma'} & \Delta[n+1] \end{array}$$

where $\sigma' : \Delta[m] \to \Delta[n+1]$ is the composite $s_{i+1} \circ \cdots \circ s_{i+k-1} \circ s_{i+k}$. Notice that $N(B)$ and $N(C)$ cover $N(P)$, and hence, since they intersect along $N(A)$:

$$N(P) = colim\big(N(B) \leftarrow N(A) \to N(C)\big)$$

We can now summarize the above discussion:

32.4. PROPOSITION. *Let $\sigma : \Delta[m] \to \Delta[n]$ be a simplex.*

1. *If σ does not contain the vertex i, then the following is a pull-back square:*

$$\begin{array}{ccc} \Delta[m] & \longrightarrow & \Delta[n+1] \\ {\scriptstyle id}\downarrow & & \downarrow{\scriptstyle s_i} \\ \Delta[m] & \xrightarrow{\sigma} & \Delta[n] \end{array}$$

2. *If the preimage $\sigma^{-1}(i)$ consists of only one element, then for some j the following is a pull-back square:*

$$\begin{array}{ccc} \Delta[m+1] & \longrightarrow & \Delta[n+1] \\ {\scriptstyle s_j}\downarrow & & \downarrow{\scriptstyle s_i} \\ \Delta[m] & \xrightarrow{\sigma} & \Delta[n] \end{array}$$

3. *Assume the preimage* $\sigma^{-1}(i)$ *consists of more than one element, i.e.,* σ *can be expressed as a composite* $\Delta[m] \xrightarrow{\sigma'} \Delta[n+1] \xrightarrow{s_i} \Delta[n]$*, where the image of* σ' *contains vertices* i *and* $i+1$*. Let* P' *be the space that fits into the pull-back square:*

$$\begin{array}{ccc} P' & \longrightarrow & \Delta[n+2] \\ \downarrow & & \downarrow{\scriptstyle s_{i+1}} \\ \Delta[m] & \xrightarrow{\sigma'} & \Delta[n+1] \end{array}$$

Then we can choose a map $\Delta[m] \to P'$ *with the following two properties. First, the composite* $\Delta[m] \to P' \to \Delta[m]$ *is the identity. Second, let us define a space* P *and a map* $P \to \Delta[m]$ *by the push-out:*

$$\begin{array}{ccc} P & = & colim\ \left(\Delta[m+1] \xleftarrow{d_{i+1}} \Delta[m] \longrightarrow P' \right) \\ \downarrow & & \searrow^{s_i} \quad \downarrow{\scriptstyle id} \quad \swarrow \\ \Delta[m] & = & \Delta[m] \end{array}$$

Then this map fits into the following pull-back square:

$$\begin{array}{ccc} P & \longrightarrow & \Delta[n+1] \\ \downarrow & & \downarrow{\scriptstyle s_i} \\ \Delta[m] & \xrightarrow{\sigma} & \Delta[n] \end{array}$$

□

33. Bounded diagrams and left Kan extensions

This section is devoted entirely to the proof of the following theorem, which is stated as Theorem 10.6 in Chapter I. It asserts that the left Kan extension preserves boundedness. The proof is based on the careful analysis of the functor ds_i we presented in the previous section.

33.1. Theorem. *Let* $f : L \to K$ *be a map of spaces. If* $F : \mathbf{L} \to \mathcal{C}$ *is a bounded diagram, then so is* $f^k F : \mathbf{K} \to \mathcal{C}$*, i.e., Kan extension along* f *induces a functor* $f^k : Fun^b(\mathbf{L}, \mathcal{C}) \to Fun^b(\mathbf{K}, \mathcal{C})$.

33.2. Lemma. *Let the following be a pull-back square of spaces:*

$$\begin{array}{ccc} P & \longrightarrow & \Delta[n+1] \\ \downarrow & & \downarrow{\scriptstyle s_i} \\ \Delta[m] & \xrightarrow{\sigma} & \Delta[n] \end{array}$$

If $F : \mathbf{\Delta}[m] \to \mathcal{C}$ *is bounded, then the induced morphism* $colim_{\mathbf{P}} F \to colim_{\mathbf{\Delta}[m]} F$ *is an isomorphism.*

Proof. If the simplex σ does not contain the vertex i, then, according to Proposition 32.4 (1), the map $P \to \Delta[m]$ coincides with $id : \Delta[m] \to \Delta[m]$. Hence in this case the lemma is obvious. We can assume therefore that σ does contain the vertex i. We can go further and assume that the map $\sigma : \Delta[m] \to \Delta[n]$ is onto. If not let $\Delta[l] \hookrightarrow \Delta[n]$ be the simplex corresponding to the image of σ. Since this simplex contains i, according to Proposition 32.4 (2), for some j, the following are pull-back squares:

$$
\begin{array}{ccccc}
P & \longrightarrow & \Delta[l+1] & \longrightarrow & \Delta[n+1] \\
\downarrow & & \downarrow{\scriptstyle s_j} & & \downarrow{\scriptstyle s_i} \\
\Delta[m] & \longrightarrow & \Delta[l] & \hookrightarrow & \Delta[n]
\end{array}
\qquad \sigma = \big(\Delta[m] \to \Delta[l] \hookrightarrow \Delta[n]\big)
$$

Thus by considering $\Delta[m] \to \Delta[l]$, instead of $\sigma : \Delta[m] \to \Delta[n]$, we can reduce the problem to the case when the map is onto.

Let $\sigma : \Delta[m] \to \Delta[n]$ be an epimorphism. We will show that, for any bounded diagram $F : \mathbf{\Delta}[m] \to \mathcal{C}$, the morphism $colim_{\mathbf{P}} F \to colim_{\mathbf{\Delta}[m]} F$ is an isomorphism. We prove it by induction on the difference $m - n$.

If $m - n = 0$, then σ corresponds to the identity $\Delta[n] \to \Delta[n]$. Hence we can identify $P \to \Delta[n]$ with $s_i : \Delta[n+1] \to \Delta[n]$. In this case the lemma follows easily from the boundedness assumption on F.

Let us assume that the statement is true for all maps $\sigma : \Delta[m] \to \Delta[n]$ where $m - n < k$. Choose a simplex $\sigma : \Delta[m] \to \Delta[n]$ for which $m - n = k$. If the preimage $\sigma^{-1}(i)$ consists of only one element, then the map $P \to \Delta[m]$ corresponds to $s_j : \Delta[m+1] \to \Delta[m]$ (see Proposition 32.4 (2)) and thus, by the boundedness assumption on F, this case is clear.

Let us assume that the preimage $\sigma^{-1}(i)$ consists of more than one element. In this case σ can be expressed as a composite $\Delta[m] \xrightarrow{\sigma'} \Delta[n+1] \xrightarrow{s_i} \Delta[n]$, where σ' is an epimorphism. Let P' be the space that fits into the following pull-back square:

$$
\begin{array}{ccc}
P' & \longrightarrow & \Delta[n+2] \\
\downarrow & & \downarrow{\scriptstyle s_{i+1}} \\
\Delta[m] & \longrightarrow & \Delta[n+1]
\end{array}
$$

According to Proposition 32.4 (3) the map $P \to \Delta[m]$ can be expressed as a push-out:

$$
\begin{array}{ccc}
P & = & colim\ \Big(\ \Delta[m+1] \xleftarrow{d_{i+1}} \Delta[m] \longrightarrow P'\ \Big) \\
\downarrow & & \searrow{\scriptstyle s_i} \quad \downarrow{\scriptstyle id} \quad \swarrow \\
\Delta[m] & = & \Delta[m]
\end{array}
$$

Hence, by applying Corollary 8.4 (1), we get:

$$
colim_{\mathbf{P}} F = colim\big(colim_{\mathbf{\Delta}[m+1]} F \leftarrow colim_{\mathbf{\Delta}[m]} F \to colim_{\mathbf{P}'} F\big)
$$

The boundedness condition on F implies that $colim_{\mathbf{\Delta}[m]} F \to colim_{\mathbf{\Delta}[m+1]} F$ is an isomorphism. It follows that so is $colim_{\mathbf{P}'} F \to colim_{\mathbf{P}} F$. By the inductive hypothesis $colim_{\mathbf{P}'} F \to colim_{\mathbf{\Delta}[m]} F$ is also an isomorphism. We can conclude that $colim_{\mathbf{P}} F \to colim_{\mathbf{\Delta}[m]} F$ is an isomorphism as well. □

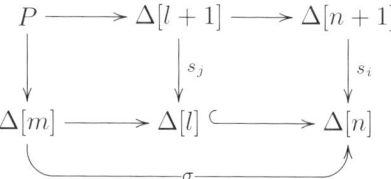

33.3. LEMMA. *Let $K \to \Delta[n]$ be a map. Consider the following pull-back square:*

$$\begin{array}{ccc} P & \longrightarrow & \Delta[n+1] \\ \downarrow & & \downarrow s_i \\ K & \longrightarrow & \Delta[n] \end{array}$$

If $F : \mathbf{K} \to \mathcal{C}$ is bounded, then the induced morphism $colim_{\mathbf{P}}F \to colim_{\mathbf{K}}F$ is an isomorphism.

PROOF. Assume first that K is finite dimensional. In this case we prove the lemma by induction on the dimension of K.

If $dim(K) = 0$, then $K = \coprod \Delta[0]$. Thus the lemma follows from Lemma 33.2 and the fact that the colimit commutes with coproducts.

Assume that the lemma is true for those spaces whose dimension is less than m. Let $dim(K) = m$. We assume for simplicity that K has only one non-degenerate simplex of dimension m, i.e., K fits into the following push-out square:

$$\begin{array}{ccccc} \partial\Delta[m] & \hookrightarrow & \Delta[m] & & \\ \downarrow & & \downarrow & & \\ L & \longrightarrow & K & \longrightarrow & \Delta[n] \end{array} \qquad \text{where } dim(L) < m.$$

The general case, when K contains more than one non-degenerate simplex of dimension m, can be proven analogously.

By pulling back $s_i : \Delta[n+1] \to \Delta[n]$ along the maps of the above diagram we get a commutative cube, where all the side squares are pull-backs and the bottom and top squares are push-outs:

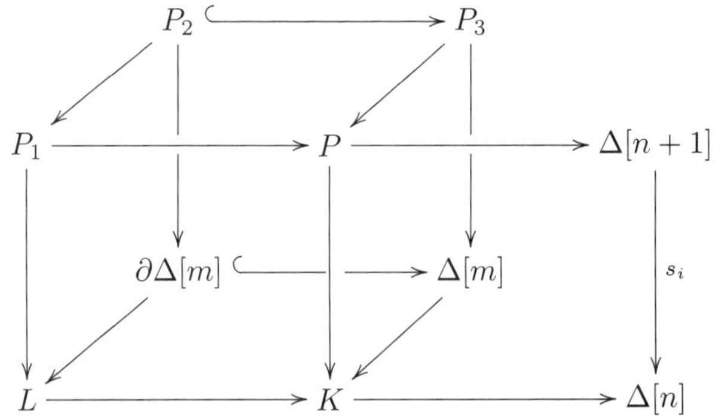

We can now apply Corollary 8.4 (1) to get the following commutative diagram:

$$\begin{array}{ccccccccc} colim_{\mathbf{P}}F & = & colim & (& colim_{\mathbf{P_1}}F & \longleftarrow & colim_{\mathbf{P_2}}F & \longrightarrow & colim_{\mathbf{P_3}}F &) \\ \downarrow & & & & \downarrow & & \downarrow & & \downarrow & \\ colim_{\mathbf{K}}F & = & colim & (& colim_{\mathbf{L}}F & \longleftarrow & colim_{\boldsymbol{\partial\Delta}[m]}F & \longrightarrow & colim_{\boldsymbol{\Delta}[m]}F &) \end{array}$$

By inductive assumption the morphisms:

$$colim_{\mathbf{P_1}}F \to colim_{\mathbf{L}}F \quad , \quad colim_{\mathbf{P_2}}F \to colim_{\boldsymbol{\partial\Delta}[m]}F$$

are isomorphisms. Moreover Lemma 33.2 implies that so is the third morphism $colim_{\mathbf{P_3}}F \to colim_{\mathbf{\Delta}[m]}F$. This shows that $colim_{\mathbf{P}}F \to colim_{\mathbf{K}}F$ is an isomorphism as well.

So far we have proven the lemma in the case when K is finite dimensional. If K is infinite dimensional, by considering its skeleton filtration and applying Corollary 8.4 (2), we can conclude that the lemma is also true in this case. □

PROOF OF THEOREM 33.1. Let $\sigma : \Delta[n] \to K$ be a simplex. Consider a degeneracy morphism $\Delta[n+1] \stackrel{s_i}{\to} \Delta[n] \stackrel{\sigma}{\to} K$. By definition the spaces $df(\sigma)$ and $df(s_i\sigma)$ fit into pull-back squares:

$$\begin{array}{ccccc} df(s_i\sigma) & \xrightarrow{df(s_i)} & df(\sigma) & \longrightarrow & L \\ \downarrow & & \downarrow & & \downarrow \\ \Delta[n+1] & \xrightarrow{s_i} & \Delta[n] & \xrightarrow{\sigma} & K \end{array}$$

If $F : \mathbf{L} \to \mathcal{C}$ is bounded, then so is the composite $\mathbf{df}(\sigma) \to \mathbf{L} \stackrel{F}{\to} \mathcal{C}$. Thus Lemma 33.3 implies that the induced morphism $colim_{\mathbf{df}(s_i\sigma)}F \to colim_{\mathbf{df}(\sigma)}F$ is an isomorphism. This shows that $f^kF : \mathbf{K} \to \mathcal{C}$ is a bounded diagram. □

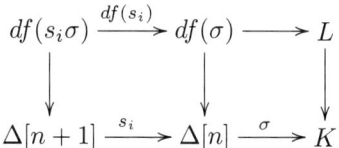

APPENDIX B

Categorical Preliminaries

34. Categories over and under an object

For reference see [**39**, Section II.6]. Let I be a small category and $\alpha : i \to j$ be a morphism in I.

By $I\downarrow i$ we denote the category whose objects are all morphisms $l \to i$, and a morphism in $I\downarrow i$ from $l_0 \to i$ to $l_1 \to i$ is a commutative triangle:

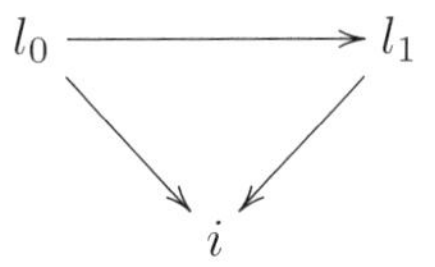

The morphism $id : i \to i$ is a terminal object in $I\downarrow i$.

By $I\downarrow \alpha$ we denote the functor $I\downarrow\alpha : I\downarrow i \to I\downarrow j$ which sends $l \to i$ to the composite $l \to i \overset{\alpha}{\to} j$. Clearly this construction defines a functor $I\downarrow - : I \to Cat$.

For any $i \in I$, there is a forgetful functor $(I\downarrow i) \to I$ which sends an object $l \to i$ in $I\downarrow i$ to l in I. These functors induce a natural transformation $(I\downarrow -) \to I$ from $I\downarrow -$ to the category I.

Dually, by $i\downarrow I$ we denote the category whose objects are all morphisms $i \to l$, and a morphism in $i\downarrow I$ from $i \to l_0$ to $i \to l_1$ is a commutative triangle:

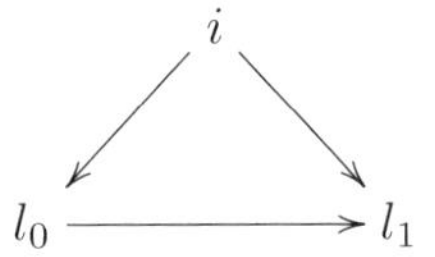

The morphism $id : i \to i$ is an initial object in $i\downarrow I$.

By $\alpha\downarrow I$ we denote the functor $\alpha\downarrow I : j\downarrow I \to i\downarrow I$ which sends the object $j \to l$ to the composite $i \overset{\alpha}{\to} j \to l$. Clearly this construction defines a functor $-\downarrow I : I^{op} \to Cat$.

35. Relative version of categories over and under an object

For reference see [**39**, Section II.6]. Let $f : J \to I$ be a functor. By $f\downarrow i$ and $i\downarrow f$ we denote the categories that fit into the following pull-back squares in Cat:

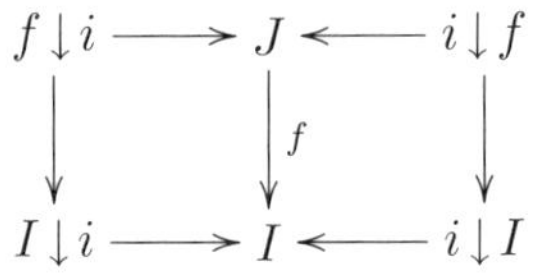

The categories $f\downarrow i$ and $i\downarrow f$ are called respectively over and under categories of f.

Explicitly, we can think about $f\downarrow i$ as a category whose objects are all the pairs $(l, f(l) \to i)$ consisting of an object l in J and a morphism $f(l) \to i$ in I. The functor $f\downarrow i \to I\downarrow i$ maps such an object to $(f(l) \to i)$. There is a similar description of $i\downarrow f$.

In the case $f = id_I$, $id\downarrow i$ and $i\downarrow id$ coincide respectively with $I\downarrow i$ and $i\downarrow I$. If $f : J \to I$ is fixed, we denote $f\downarrow i$ and $i\downarrow f$ simply by $J\downarrow i$ and $i\downarrow J$.

36. Pull-back process and Kan extensions

For reference see [**39**, Section X]. Let $\mathcal{C}$ be a category closed under colimits. Consider a functor $f : J \to I$. With f we can associate two other functors. The pull-back process $f^* : Fun(I,\mathcal{C}) \to Fun(J,\mathcal{C})$ which assigns to a functor $H : I \to \mathcal{C}$ the composite $J \xrightarrow{f} I \xrightarrow{H} \mathcal{C}$. The left Kan extension $f^k : Fun(J,\mathcal{C}) \to Fun(I,\mathcal{C})$, which is left adjoint to the pull-back process. It can be constructed explicitly as follows. Let $H : J \to \mathcal{C}$ be a functor. For every $i \in I$ pull-back H along $f\downarrow i \to J$ and take the colimit $colim_{f\downarrow i}H$. In this way we get a functor $colim_{f\downarrow -}H : I \to \mathcal{C}$. The assignment:

$$Fun(J,\mathcal{C}) \ni H \mapsto colim_{f\downarrow -}H \in Fun(I,\mathcal{C})$$

is natural in H, and hence it induces a functor $Fun(J,\mathcal{C}) \to Fun(I,\mathcal{C})$. This functor is called the left Kan extension along f.

One defines dually the right Kan extension $f_k : Fun(J,\mathcal{C}) \to Fun(I,\mathcal{C})$, which is right adjoint to the pull-back process. It assigns to $H : J \to \mathcal{C}$, the functor $(f_k H)(i) := lim_{i\downarrow f}H$.

36.1. EXAMPLE. The left Kan extension along $I \to *$ is the colimit functor $colim_I : Fun(I,\mathcal{C}) \to \mathcal{C}$.

The left Kan extension process commutes with composition of functors and hence does not modify colimits. Explicitly, the left Kan extension enjoys the following two properties, which are straightforward consequences of the definition:

36.2. PROPOSITION. *Let $\mathcal{C}$ be a category closed under colimits.*

1. *Let $f : J \to I$ and $g : I \to L$ be functors of small categories. Then $(g \circ f)^k$ can be identified with $g^k \circ f^k$.*
2. *For any functor $f : J \to I$, the following diagram commutes:*

$$\begin{array}{ccc} Fun(J,\mathcal{C}) & \xrightarrow{\quad f^k \quad} & Fun(I,\mathcal{C}) \\ & {\scriptstyle colim_J} \searrow \quad \swarrow {\scriptstyle colim_I} & \\ & \mathcal{C} & \end{array}$$

□

37. Cofinality for colimits

For reference see [**39**, Theorem IX.3.1]. Let $\mathcal{C}$ be a category closed under colimits. The main application of relative categories under an object is in cofinality statements for colimits. We say that a functor $f : J \to I$ is *cofinal* with respect to taking colimits if, for any $F : I \to \mathcal{C}$, the induced morphism $colim_J f^*F \to colim_I F$ is an isomorphism.

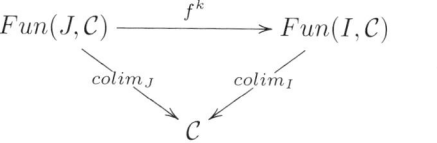

37.1. PROPOSITION. *Let $f : J \to I$ be a functor of small categories. If for any $i \in I$, the under category $i \downarrow f$ is non-empty and connected (the nerve $N(i \downarrow f)$ is a non-empty and connected space), then $f : J \to I$ is cofinal with respect to taking colimits.* □

38. Grothendieck construction

For reference see [**47**, Definition 1.1]. Let $H : I \to Cat$ be a functor. The *Grothendieck construction* on H is the category $Gr_I H$, where:

- an object in $Gr_I H$ is a pair (i, a) consisting of an object $i \in I$ and an object $a \in H(i)$;
- a morphism $(\alpha, h) : (i, a) \to (j, b)$ in $Gr_I H$ is a pair (α, h) consisting of a morphism $\alpha : i \to j$ in I and a morphism $h : H(\alpha)(a) \to b$ in $H(j)$;
- the composition of $(\alpha, h) : (i, a) \to (j, b)$ and $(\beta, g) : (j, b) \to (l, c)$ is defined to be $\big(\beta \circ \alpha, g \circ H(\beta)(h)\big) : (i, a) \to (l, c)$.

The construction $Gr_I H$ is natural. Let $f : J \to I$ be a functor of small categories, $H_1 : J \to Cat$ and $H_0 : I \to Cat$ be functors, and $\Psi : H_1 \to f^* H_0$ be a natural transformation. This data induces a functor $Gr_f \Psi : Gr_J H_1 \to Gr_I H_0$ which sends an object (j, a) to $\big(f(j), \Psi_j(a)\big)$.

38.1. EXAMPLE. If $H : I \to Cat$ is the constant functor with value J, then the Grothendieck construction $Gr_I H$ is isomorphic to the product category $I \times J$.

38.2. EXAMPLE. Let us consider the category given by the following graph:

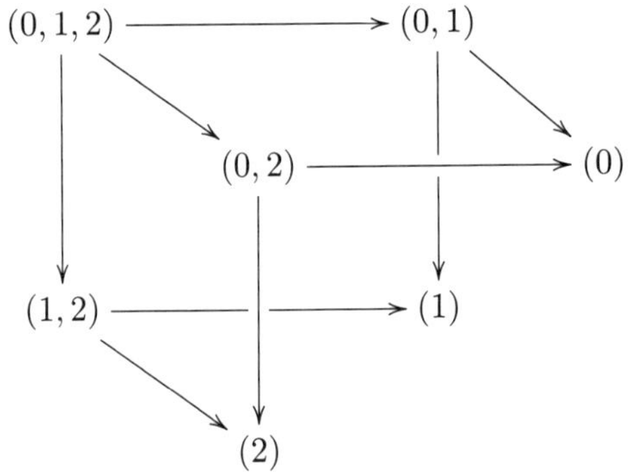

This category is isomorphic to the following Grothendieck construction:

$$Gr\left((2) \leftarrow \begin{pmatrix} (0,1,2) \longrightarrow (0,2) \\ \downarrow \\ (1,2) \end{pmatrix} \to \begin{pmatrix} (0,1) \longrightarrow (0) \\ \downarrow \\ (1) \end{pmatrix}\right)$$

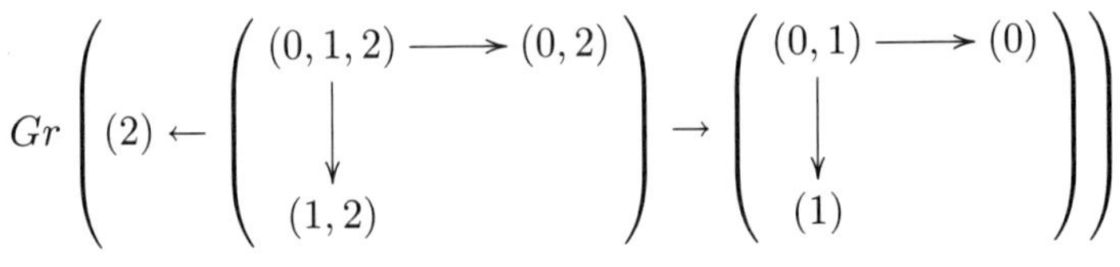

39. Grothendieck construction & the pull-back process

The category $Gr_I H$ is equipped with a functor $Gr_I H \to I$, $(i, a) \mapsto i$. This functor behaves well with regard to taking pull-backs in Cat. Let $f : J \to I$ be a functor and $i \in I$ be an object. The following are pull-back squares in Cat:

$$\begin{array}{ccccc} Gr_J H & \longrightarrow & Gr_I H & \longleftarrow & H(i) \\ \downarrow & & \downarrow & & \downarrow \\ J & \xrightarrow{f} & I & \longleftarrow & \{i\} \end{array}$$

In particular, by pulling back $Gr_I H \to I$ along $I \downarrow i \to I$ we get a natural isomorphism $Gr_{I\downarrow i} H \cong (Gr_I H) \downarrow i$.

40. Functors indexed by Grothendieck constructions

Here we present two ways of describing functors indexed by a Grothendieck construction. We start with the less economical.

40.1. DEFINITION. Let $H : I \to Cat$ be a functor. We say that a family of functors $F = \{F_i : I\downarrow i \to Fun\big(H(i), \mathcal{C}\big)\}_{i\in I}$ is *compatible* over H if for any morphism $\alpha : j \to i$ in I the following diagram commutes:

$$\begin{array}{ccc} I\downarrow j & \xrightarrow{F_j} & Fun\big(H(j), \mathcal{C}\big) \\ {\scriptstyle I\downarrow\alpha}\downarrow & & \downarrow{\scriptstyle H(\alpha)^k} \\ I\downarrow i & \xrightarrow{F_i} & Fun\big(H(i), \mathcal{C}\big) \end{array}$$

We say that a family of natural transformations $\Psi = \{\Psi_i : F_i \to G_i\}_{i\in I}$ is a morphism between compatible families F and G over H, if for any $\alpha : j \to i$ in I, the pull-back $(I\downarrow\alpha)^*\Psi_i$ coincides with $H(\alpha)^k\Psi_j$.

Compatible families over H together with morphisms, as defined above, clearly form a category.

With a compatible family $F = \{F_i\}_{i\in I}$ we can associate a functor which is denoted by the same symbol $F : Gr_I H \to \mathcal{C}$. It assigns to $(i, a) \in Gr_I H$ the object $F_i(i \xrightarrow{id} i) \in \mathcal{C}$. Conversely, to a functor $F : Gr_I H \to \mathcal{C}$ we can associate a compatible family $\{F_i\}_{i\in I}$, where $F_i : I\downarrow i \to Fun\big(H(i), \mathcal{C}\big)$ assigns to $\alpha : j \to i$ the functor $H(i) \ni a \mapsto \big(H(\alpha)^k F(j, -)\big)(a) \in \mathcal{C}$. It is not difficult to see that in this way we get inverse isomorphisms between the category of compatible families over H and the functor category $Fun(Gr_I H, \mathcal{C})$. Thus we do not distinguish between those two categories and we use the symbol $Fun(Gr_I H, \mathcal{C})$ to denote both of them. We sometimes refer to the compatible family associated to a functor $F : Gr_I H \to \mathcal{C}$ as the *local presentation* of F.

A compatible family over H carries a lot of redundant data just to describe a functor indexed by $Gr_I H$. We can be more efficient. To describe a functor $F : Gr_I H \to \mathcal{C}$ it is necessary and sufficient to have the following data:

1. a functor $F_i : H(i) \to \mathcal{C}$ for every object i in I;
2. a natural transformation $F_\alpha : F_j \to H(\alpha)^* F_i$ for every morphism $\alpha : j \to i$ in I;
3. if $\alpha : j \to i$ and $\beta : i \to l$ are composable morphisms in I, then the natural transformations $F_{\beta\circ\alpha}$ and $\big(H(\alpha)^* F_\beta\big) \circ F_\alpha$ should coincide.

Let $F : Gr_I H \to \mathcal{C}$ be a diagram. The functor $F_i : H(i) \to \mathcal{C}$ is given by the composite $H(i) \hookrightarrow Gr_I H \xrightarrow{F} \mathcal{C}$. The natural transformation $F_\alpha : F_j \to H(\alpha)^* F_i$ is induced by the morphisms $F(\alpha, id) : F(j, a) \to F\big(j, H(\alpha)(a)\big)$.

By applying the colimit to this data we get a functor:

$$colim_{H(-)}F : I \to \mathcal{C}\ ,\ i \mapsto colim_{H(i)}F\ ,\ (\alpha : i \to j) \mapsto colim\, F_\alpha$$

Let $colim_{H(-)}F \to colim_{Gr_I H}F$ be the natural transformation induced by the morphisms $colim_{H(i)}F \to colim_{Gr_I H}F$.

40.2. PROPOSITION. *The natural transformation* $colim_{H(-)}F \to colim_{Gr_I H}F$ *satisfies the universal property of the colimit of the diagram* $colim_{H(-)}F : I \to \mathcal{C}$. *The induced morphism* $colim_{i\in I} colim_{H(i)}F \to colim_{Gr_I H}F$ *is therefore an isomorphism.* □

Bibliography

[1] D. W. Anderson. Fibrations and geometric realizations. *Bull. Amer. Math. Soc.*, 84(5):765–788, 1978.

[2] A. Bak and C. Weibel. A tribute to Robert Wayne Thomason (1952–1995). *K-Theory*, 12(1):1–2, 1997.

[3] A. K. Bousfield. The localization of spaces with respect to homology. *Topology*, 14:133–150, 1975.

[4] A. K. Bousfield. The localization of spectra with respect to homology. *Topology*, 18(4):257–281, 1979.

[5] A. K. Bousfield. Localization and periodicity in unstable homotopy theory. *J. Amer. Math. Soc.*, 7(4):831–873, 1994.

[6] A. K. Bousfield and E. M. Friedlander. Homotopy theory of γ-spaces, spectra, and bisimplicial sets. In *Geometric applications of homotopy theory (Proc. Conf., Evanston, Ill., 1977), II*, volume 658 of *Lecture Notes in Math.*, pages 80–130. Springer, Berlin, 1978.

[7] A. K. Bousfield and D. M. Kan. *Homotopy limits, completions and localizations*, volume 304 of *Lecture Notes in Mathematics*. Springer-Verlag, Berlin, 1972.

[8] K. S. Brown. Abstract homotopy theory and generalized sheaf cohomology. *Trans. Amer. Math. Soc.*, 186:419–458, 1974.

[9] C. Casacuberta. Recent advances in unstable localization. In *The Hilton Symposium 1993 (Montreal, PQ)*, volume 6 of *CRM Proc. Lecture Notes*, pages 1–22. Amer. Math. Soc., Providence, RI, 1994.

[10] J.-M. Cordier and T. Porter. Vogt's theorem on categories of homotopy coherent diagrams. *Math. Proc. Cambridge Philos. Soc.*, 100(1):65–90, 1986.

[11] J.-M. Cordier and T. Porter. Homotopy coherent category theory. *Trans. Amer. Math. Soc.*, 349(1):1–54, 1997.

[12] E. B. Curtis. Simplicial homotopy theory. *Advances in Math.*, 6:107–209 (1971), 1971.

[13] E. S. Devinatz, M. J. Hopkins, and J. H. Smith. Nilpotence and stable homotopy theory. I. *Ann. of Math. (2)*, 128(2):207–241, 1988.

[14] E. Dror Farjoun. Homotopy and homology of diagrams of spaces. In *Algebraic topology (Seattle, Wash., 1985)*, volume 1286 of *Lecture Notes in Math.*, pages 93–134. Springer, Berlin, 1987.

[15] E. Dror Farjoun. Homotopy theories for diagrams of spaces. *Proc. Amer. Math. Soc.*, 101(1):181–189, 1987.

[16] E. Dror Farjoun and A. Zabrodsky. Homotopy equivalence between diagrams of spaces. *J. Pure Appl. Algebra*, 41(2-3):169–182, 1986.

[17] W. G. Dwyer. Homology decompositions for classifying spaces of finite groups. *Topology*, 36(4):783–804, 1997.

[18] W. G. Dwyer, P. S. Hirschhorn, and D. M. Kan. *Model categories and more general abstract homotopy theory: a work in what we like to think of as a progress (title as of March, 1997)*. Preprint, available at: `http://www-math.mit.edu/~psh/`.

[19] W. G. Dwyer and D. M. Kan. Calculating simplicial localizations. *J. Pure Appl. Algebra*, 18(1):17–35, 1980.

[20] W. G. Dwyer and D. M. Kan. Simplicial localizations of categories. *J. Pure Appl. Algebra*, 17(3):267–284, 1980.

[21] W. G. Dwyer and D. M. Kan. Function complexes for diagrams of simplicial sets. *Nederl. Akad. Wetensch. Indag. Math.*, 45(2):139–147, 1983.

[22] W. G. Dwyer and D. M. Kan. A classification theorem for diagrams of simplicial sets. *Topology*, 23(2):139–155, 1984.

[23] W. G. Dwyer and D. M. Kan. An obstruction theory for diagrams of simplicial sets. *Nederl. Akad. Wetensch. Indag. Math.*, 46(2):139–146, 1984.

[24] W. G. Dwyer and D. M. Kan. Realizing diagrams in the homotopy category by means of diagrams of simplicial sets. *Proc. Amer. Math. Soc.*, 91(3):456–460, 1984.

[25] W. G. Dwyer and D. M. Kan. Singular functors and realization functors. *Nederl. Akad. Wetensch. Indag. Math.*, 46(2):147–153, 1984.

[26] W. G. Dwyer and D. M. Kan. Equivariant homotopy classification. *J. Pure Appl. Algebra*, 35(3):269–285, 1985.

[27] W. G. Dwyer and D. M. Kan. Reducing equivariant homotopy theory to the theory of fibrations. In *Conference on algebraic topology in honor of Peter Hilton (Saint John's, Nfld., 1983)*, volume 37 of *Contemp. Math.*, pages 35–49. Amer. Math. Soc., Providence, R.I., 1985.

[28] W. G. Dwyer and D. M. Kan. Equivalences between homotopy theories of diagrams. In *Algebraic topology and algebraic K-theory (Princeton, N.J., 1983)*, volume 113 of *Ann. of Math. Stud.*, pages 180–205. Princeton Univ. Press, Princeton, NJ, 1987.

[29] W. G. Dwyer and D. M. Kan. Centric maps and realization of diagrams in the homotopy category. *Proc. Amer. Math. Soc.*, 114(2):575–584, 1992.

[30] W. G. Dwyer, H. R. Miller, and C. W. Wilkerson. The homotopic uniqueness of BS^3. In *Algebraic topology, Barcelona, 1986*, pages 90–105. Springer, Berlin, 1987.

[31] W. G. Dwyer and J. Spaliński. Homotopy theories and model categories. In *Handbook of algebraic topology*, pages 73–126. North-Holland, Amsterdam, 1995.

[32] D. A. Edwards and H. M. Hastings. *Čech and Steenrod homotopy theories with applications to geometric topology*, volume 542 of *Lecture Notes in Mathematics*. Springer-Verlag, Berlin-New York, 1972.

[33] E. D. Farjoun. *Cellular spaces, null spaces and homotopy localization*, volume 1622 of *Lecture Notes in Mathematics*. Springer-Verlag, Berlin, 1996.

[34] P. S. Hirschhorn. Localization of model categories. Unpublished, available at P. Hirschhorn's homepage: `http://www-math.mit.edu/~psh/`.

[35] M. Hovey. *Model categories*. Mathematical Surveys and Monographs, 63. American Mathematical Society, Providence, RI, 1999.

[36] S. Jackowski, J. McClure, and R. Oliver. Homotopy classification of self-maps of BG via G-actions. I. *Ann. of Math. (2)*, 135(1):183–226, 1992.

[37] S. Jackowski, J. McClure, and R. Oliver. Homotopy classification of self-maps of BG via G-actions. II. *Ann. of Math. (2)*, 135(2):227–270, 1992.

[38] D. M. Latch, R. W. Thomason, and W. S. Wilson. Simplicial sets from categories. *Math. Z.*, 164(3):195–214, 1979.

[39] S. Mac Lane. *Categories for the working mathematician*, volume 5 of *Graduate Texts in Mathematics*. Springer-Verlag, New York, 1971.

[40] J. P. May. *Simplicial objects in algebraic topology*. Chicago Lectures in Mathematics. University of Chicago Press, Chicago, IL, 1992. Reprint of the 1967 original.

[41] D. Quillen. *Homotopical algebra*, volume 43 of *Lecture Notes in Mathematics*. Springer-Verlag, Berlin, 1967.

[42] D. Quillen. Rational homotopy theory. *Ann. of Math. (2)*, 90:205–295, 1969.

[43] D. Quillen. Homotopy properties of the poset of nontrivial p-subgroups of a group. *Adv. in Math.*, 28(2):101–128, 1978.

[44] C. I. Reedy. Homotopy theory of model categories. Unpublished, available at P. Hirschhorn's homepage: `http://www-math.mit.edu/~psh/`.

[45] G. Segal. Classifying spaces and spectral sequences. *Inst. Hautes Études Sci. Publ. Math. No.*, 34:105–112, 1968.

[46] J. Sł omińska. Homotopy colimits on e-i-categories. In *Algebraic topology Poznań 1989*, volume 1474 of *Lecture Notes in Math.*, pages 273–294. Springer, Berlin, 1991.

[47] R. W. Thomason. Homotopy colimits in the category of small categories. *Math. Proc. Cambridge Philos. Soc.*, 85(1):91–109, 1979.

[48] R. W. Thomason. Cat as a closed model category. *Cahiers Topologie Géom. Différentielle*, 21(3):305–324, 1980.

[49] R. W. Thomason. First quadrant spectral sequences in algebraic K-theory via homotopy colimits. *Comm. Algebra*, 10(15):1589–1668, 1982.

[50] R. M. Vogt. Homotopy limits and colimits. *Math. Z.*, 134:11–52, 1973.

Index

Editorial Information

To be published in the *Memoirs*, a paper must be correct, new, nontrivial, and significant. Further, it must be well written and of interest to a substantial number of mathematicians. Piecemeal results, such as an inconclusive step toward an unproved major theorem or a minor variation on a known result, are in general not acceptable for publication. Papers appearing in *Memoirs* are generally longer than those appearing in *Transactions*, which shares the same editorial committee.

As of September 30, 2001, the backlog for this journal was approximately 7 volumes. This estimate is the result of dividing the number of manuscripts for this journal in the Providence office that have not yet gone to the printer on the above date by the average number of monographs per volume over the previous twelve months, reduced by the number of volumes published in four months (the time necessary for preparing a volume for the printer). (There are 6 volumes per year, each containing at least 4 numbers.)

A Consent to Publish and Copyright Agreement is required before a paper will be published in the *Memoirs*. After a paper is accepted for publication, the Providence office will send a Consent to Publish and Copyright Agreement to all authors of the paper. By submitting a paper to the *Memoirs*, authors certify that the results have not been submitted to nor are they under consideration for publication by another journal, conference proceedings, or similar publication.

Information for Authors

Memoirs are printed from camera copy fully prepared by the author. This means that the finished book will look exactly like the copy submitted.

The paper must contain a *descriptive title* and an *abstract* that summarizes the article in language suitable for workers in the general field (algebra, analysis, etc.). The *descriptive title* should be short, but informative; useless or vague phrases such as "some remarks about" or "concerning" should be avoided. The *abstract* should be at least one complete sentence, and at most 300 words. Included with the footnotes to the paper should be the 2000 *Mathematics Subject Classification* representing the primary and secondary subjects of the article. The classifications are accessible from `www.ams.org/msc/`. The list of classifications is also available in print starting with the 1999 annual index of *Mathematical Reviews*. The Mathematics Subject Classification footnote may be followed by a list of *key words and phrases* describing the subject matter of the article and taken from it. Journal abbreviations used in bibliographies are listed in the latest *Mathematical Reviews* annual index. The series abbreviations are also accessible from `www.ams.org/publications/`. To help in preparing and verifying references, the AMS offers MR Lookup, a Reference Tool for Linking, at `www.ams.org/mrlookup/`. When the manuscript is submitted, authors should supply the editor with electronic addresses if available. These will be printed after the postal address at the end of the article.

Electronically prepared manuscripts. The AMS encourages electronically prepared manuscripts, with a strong preference for $\mathcal{A}_{\mathcal{M}}\mathcal{S}$-LaTeX. To this end, the Society has prepared $\mathcal{A}_{\mathcal{M}}\mathcal{S}$-LaTeX author packages for each AMS publication. Author packages include instructions for preparing electronic manuscripts, the *AMS Author Handbook*, samples, and a style file that generates the particular design specifications of that publication series. Though $\mathcal{A}_{\mathcal{M}}\mathcal{S}$-LaTeX is the highly preferred format of TeX, author packages are also available in $\mathcal{A}_{\mathcal{M}}\mathcal{S}$-TeX.

Authors may retrieve an author package from e-MATH starting from `www.ams.org/tex/` or via FTP to `ftp.ams.org` (login as `anonymous`, enter username as password, and type `cd pub/author-info`). The *AMS Author Handbook* and the *Instruction Manual* are available in PDF format following the author packages link from `www.ams.org/tex/`. The author package can be obtained free of charge by sending email to `pub@ams.org` (Internet) or from the Publication Division, American Mathematical Society, P.O. Box 6248, Providence, RI 02940-6248. When requesting an author package, please specify AMS-LaTeX or AMS-TeX, Macintosh or IBM (3.5) format, and the publication in which your paper will appear. Please be sure to include your complete mailing address.

Sending electronic files. After acceptance, the source file(s) should be sent to the Providence office (this includes any TeX source file, any graphics files, and the DVI or PostScript file).

Before sending the source file, be sure you have proofread your paper carefully. The files you send must be the EXACT files used to generate the proof copy that was accepted for publication. For all publications, authors are required to send a printed copy of their paper, which exactly matches the copy approved for publication, along with any graphics that will appear in the paper.

TeX files may be submitted by email, FTP, or on diskette. The DVI file(s) and PostScript files should be submitted only by FTP or on diskette unless they are encoded properly to submit through email. (DVI files are binary and PostScript files tend to be very large.)

Electronically prepared manuscripts can be sent via email to `pub-submit@ams.org` (Internet). The subject line of the message should include the publication code to identify it as a Memoir. TeX source files, DVI files, and PostScript files can be transferred over the Internet by FTP to the Internet node `e-math.ams.org` (130.44.1.100).

Electronic graphics. Comprehensive instructions on preparing graphics are available at `www.ams.org/jourhtml/graphics.html`. A few of the major requirements are given here.

Submit files for graphics as EPS (Encapsulated PostScript) files. This includes graphics originated via a graphics application as well as scanned photographs or other computer-generated images. If this is not possible, TIFF files are acceptable as long as they can be opened in Adobe Photoshop or Illustrator. No matter what method was used to produce the graphic, it is necessary to provide a paper copy to the AMS.

Authors using graphics packages for the creation of electronic art should also avoid the use of any lines thinner than 0.5 points in width. Many graphics packages allow the user to specify a "hairline" for a very thin line. Hairlines often look acceptable when proofed on a typical laser printer. However, when produced on a high-resolution laser imagesetter, hairlines become nearly invisible and will be lost entirely in the final printing process.

Screens should be set to values between 15% and 85%. Screens which fall outside of this range are too light or too dark to print correctly. Variations of screens within a graphic should be no less than 10%.

Inquiries. Any inquiries concerning a paper that has been accepted for publication should be sent directly to the Electronic Prepress Department, American Mathematical Society, P. O. Box 6248, Providence, RI 02940-6248.

Editors

This journal is designed particularly for long research papers, normally at least 80 pages in length, and groups of cognate papers in pure and applied mathematics. Papers intended for publication in the *Memoirs* should be addressed to one of the following editors. In principle the Memoirs welcomes electronic submissions, and some of the editors, those whose names appear below with an asterisk (*), have indicated that they prefer them. However, editors reserve the right to request hard copies after papers have been submitted electronically. Authors are advised to make preliminary email inquiries to editors about whether they are likely to be able to handle submissions in a particular electronic form.

Algebra to KAREN E. SMITH, Department of Mathematics, University of Michigan, 525 University, Suite 2832, Ann Arbor, MI 48109-1109; email: `kesmith@math.lsa.umich.edu`

Algebraic geometry and commutative algebra to LAWRENCE EIN, Department of Mathematics, University of Illinois, 851 S. Morgan (M/C 249), Chicago, IL 60607-7045; email: `ein@uic.edu`

Algebraic topology and cohomology of groups to STEWART PRIDDY, Department of Mathematics, Northwestern University, 2033 Sheridan Road, Evanston, IL 60208-2730; email: `priddy@math.nwu.edu`

Combinatorics and Lie theory to SERGEY FOMIN, Department of Mathematics, University of Michigan, Ann Arbor, Michigan 48109-1109; email: `fomin@math.lsa.umich.edu`

Complex analysis and complex geometry to DUONG H. PHONG, Department of Mathematics, Columbia University, 2990 Broadway, New York, NY 10027-0029; email: `phong@math.columbia.edu`

***Differential geometry and global analysis** to LISA C. JEFFREY, Department of Mathematics, University of Toronto, 100 St. George St., Toronto, ON Canada M5S 3G3; email: `jeffrey@math.toronto.edu`

***Dynamical systems and ergodic theory** to ROBERT F. WILLIAMS, Department of Mathematics, University of Texas, Austin, Texas 78712-1082; email: `bob@math.utexas.edu`

Functional analysis and operator algebras to DAN VOICULESCU, Department of Mathematics, University of California, Berkeley, 970 Evans Hall, Floor 9, Berkeley, CA 94720-0001; email: `dvv@math.berkeley.edu`

Geometric topology, knot theory and hyperbolic geometry to ABIGAIL A. THOMPSON, Department of Mathematics, University of California, Davis, Davis, CA 95616-5224; email: `thompson@math.ucdavis.edu`

Harmonic analysis, representation theory, and Lie theory to ROBERT J. STANTON, Department of Mathematics, The Ohio State University, 231 West 18th Avenue, Columbus, OH 43210-1174; email: `stanton@math.ohio-state.edu`

***Logic** to THEODORE SLAMAN, Department of Mathematics, University of California, Berkeley, CA 94720-3840; email: `slaman@math.berkeley.edu`

Number theory to HAROLD G. DIAMOND, Department of Mathematics, University of Illinois, 1409 W. Green St., Urbana, IL 61801-2917; email: `diamond@math.uiuc.edu`

***Ordinary differential equations, partial differential equations, and applied mathematics** to PETER W. BATES, Department of Mathematics, Brigham Young University, 292 TMCB, Provo, UT 84602-1001; email: `peter@math.byu.edu`

***Probability and statistics** to KRZYSZTOF BURDZY, Department of Mathematics, University of Washington, Box 354350, Seattle, Washington 98195-4350; email: `burdzy@math.washington.edu`

***Real and harmonic analysis and geometric partial differential equations** to WILLIAM BECKNER, Department of Mathematics, University of Texas, Austin, TX 78712-1082; email: `beckner@math.utexas.edu`

All other communications to the editors should be addressed to the Managing Editor, WILLIAM BECKNER, Department of Mathematics, University of Texas, Austin, TX 78712-1082; email: `beckner@math.utexas.edu`.

Selected Titles in This Series

(Continued from the front of this publication)

705 **I. Moerdijk and J. J. C. Vermeulen,** Proper maps of toposes, 2000

704 **Jeff Hooper, Victor Snaith, and Min van Tran,** The second Chinburg conjecture for quaternion fields, 2000

703 **Erik Guentner, Nigel Higson, and Jody Trout,** Equivariant E-theory for C^*-algebras, 2000

702 **Ilijas Farah,** Analytic guotients: Theory of liftings for quotients over analytic ideals on the integers, 2000

701 **Paul Selick and Jie Wu,** On natural coalgebra decompositions of tensor algebras and loop suspensions, 2000

700 **Vicente Cortés,** A new construction of homogeneous quaternionic manifolds and related geometric structures, 2000

699 **Alexander Fel′shtyn,** Dynamical zeta functions, Nielsen theory and Reidemeister torsion, 2000

698 **Andrew R. Kustin,** Complexes associated to two vectors and a rectangular matrix, 2000

697 **Deguang Han and David R. Larson,** Frames, bases and group representations, 2000

696 **Donald J. Estep, Mats G. Larson, and Roy D. Williams,** Estimating the error of numerical solutions of systems of reaction-diffusion equations, 2000

695 **Vitaly Bergelson and Randall McCutcheon,** An ergodic IP polynomial Szemerédi theorem, 2000

694 **Alberto Bressan, Graziano Crasta, and Benedetto Piccoli,** Well-posedness of the Cauchy problem for $n \times n$ systems of conservation laws, 2000

693 **Doug Pickrell,** Invariant measures for unitary groups associated to Kac-Moody Lie algebras, 2000

692 **Mara D. Neusel,** Inverse invariant theory and Steenrod operations, 2000

691 **Bruce Hughes and Stratos Prassidis,** Control and relaxation over the circle, 2000

690 **Robert Rumely, Chi Fong Lau, and Robert Varley,** Existence of the sectional capacity, 2000

689 **M. A. Dickmann and F. Miraglia,** Special groups: Boolean-theoretic methods in the theory of quadratic forms, 2000

688 **Piotr Hajłasz and Pekka Koskela,** Sobolev met Poincaré, 2000

687 **Guy David and Stephen Semmes,** Uniform rectifiability and quasiminimizing sets of arbitrary codimension, 2000

686 **L. Gaunce Lewis, Jr.,** Splitting theorems for certain equivariant spectra, 2000

685 **Jean-Luc Joly, Guy Metivier, and Jeffrey Rauch,** Caustics for dissipative semilinear oscillations, 2000

684 **Harvey I. Blau, Bangteng Xu, Z. Arad, E. Fisman, V. Miloslavsky, and M. Muzychuk,** Homogeneous integral table algebras of degree three: A trilogy, 2000

683 **Serge Bouc,** Non-additive exact functors and tensor induction for Mackey functors, 2000

682 **Martin Majewski,** ational homotopical models and uniqueness, 2000

681 **David P. Blecher, Paul S. Muhly, and Vern I. Paulsen,** Categories of operator modules (Morita equivalence and projective modules, 2000

680 **Joachim Zacharias,** Continuous tensor products and Arveson's spectral C^*-algebras, 2000

679 **Y. A. Abramovich and A. K. Kitover,** Inverses of disjointness preserving operators, 2000